Beiträge zur organischen Synthese

Band 102

T0134339

Beiträge zur organischen Synthese

Band 102

Herausgegeben von
Prof. Dr. Stefan Bräse

Karlsruher Institut für Technologie (KIT)
Institut für Organische Chemie
Fritz-Haber-Weg 6, D-76131 Karlsruhe

Institut für Biologische und Chemische Systeme – Funktionelle Molekulare Systeme
Hermann-von-Helmholtz-Platz 1
D-76344 Eggenstein-Leopoldshafen

Steffen Andreas Otterbach

Organic Semiconductors based on [2.2]Paracyclophanes and Porphyrins as Hole Transport Materials in Perovskite Solar Cells

Logos Verlag Berlin

 λογος

Bibliographic information published by the Deutsche Nationalbibliothek

The Deutsche Nationalbibliothek lists this publication in the Deutsche Nationalbibliografie; detailed bibliographic data are available in the Internet at http://dnb.d-nb.de.

ISBN 978-3-8325-5709-6
ISSN 1862-5681

Logos Verlag Berlin GmbH
Georg-Knorr-Str. 4, Geb. 10, 12681 Berlin

Tel.:   +49 (0)30 / 42 85 10 90
Fax: +49 (0)30 / 42 85 10 92

https://www.logos-verlag.de

Organische Halbleiter auf Basis von [2.2]Paracyclophanen und Porphyrinen als Lochtransportschichten in Perowskit Solarzellen

TABLE OF CONTENTS

German Title of the Thesis...I

Table of Contents III

Abstract 1

Inhaltliche Zusammenfassung.. 3

1   Introduction.. 5

  1.1   The Functionality of Solar Cells 7

  1.2   History of Solar Cells – The First Generation.......................... 11

  1.3   The Second Generation – Thin-Film Solar Cells 12

  1.4   The Third Generation – Emerging Solar Cells........................ 12

    1.4.1   Organic Solar Cells (OSCs).. 12

    1.4.2   Dye-Sensitized Solar Cells (DSSCs) 13

    1.4.3   Perovskite Solar Cells (PSCs)... 14

  1.5   Charge Transport Materials – Hole Transport Materials (HTMs)....... 16

    1.5.1   Electron Transport Materials (ETMs)............................... 16

    1.5.2   Inorganic Hole Transport Materials................................. 17

    1.5.3   Organic Hole Transport Materials 18

2   Aim of the work.. 33

3   Main Section 35

  3.1   Syntheses of the Donor Moieties.. 35

    3.1.1   Synthesis of the substituted Triarylamines 36

    3.1.2   Synthesis of the substituted Carbazoles 38

3.2   Pseudo-*para*-substituted [2.2]Paracyclophanes 39

   3.2.1   Synthesis and Characterization 40

   3.2.2   Optical and Electrochemical Characterization 55

   3.2.3   Applicability of disubstituted [2.2]Paracyclophanes 66

3.3   Bis(pseudo-*meta*)-*para*-substituted [2.2]Paracyclophanes 70

   3.3.1   Synthesis and Characterization 70

   3.3.2   Optical and Electrochemical Characterization 78

   3.3.3   Applicability of tetra-substituted [2.2]Paracyclophanes 85

3.4   *Meso*-tetrakis-substituted Porphyrins 86

   3.4.1   Synthesis and Characterization 87

   3.4.2   Optical and Electrochemical Characterization 95

   3.4.3   Applicability of tetra-substituted Porphyrins 105

3.5   Tetra-*para*-substituted Tetraphenylethenes 107

   3.5.1   Synthesis and Characterization 108

   3.5.2   Optical and Electrochemical Characterization 111

3.6   Manufactured Devices 116

   3.6.1   Pseudo-*para*-substituted [2.2]Paracyclophanes 116

   3.6.2   Bis(pseudo-*meta*)-*para*-substituted [2.2]Paracyclophanes 118

4   Conclusion and Outlook 121

5   Experimental Section 125

5.1   General Remarks 125

   5.1.1   Materials and Methods 125

5.1.2   Devices 126

5.2   Synthesis and Characterization of the Synthesized Compounds 129

5.2.1   Synthesis of Donor Moieties 129

5.2.2   Pseudo-*para*-substituted [2.2]Paracyclophancs 139

5.2.3   Bis(pseudo-*meta*)-*para*-substituted [2.2]Paracyclophanes 172

5.2.4   *Meso*-tetrakis-substituted Porphyrins 180

5.2.5   Tetra-*para*-substituted Tetraphenylethenes 207

6   Abbreviations 221

7   Bibliography 225

8   Appendix i

8.1   Supplementary Data i

Thermogravimetric analysis i

8.2   Curriculum Vitae iii

Education iii

Achievements iv

8.3   List of Publications iv

8.4   Acknowledgments v

ABSTRACT

Climate change and the finiteness of fossil fuels call for the development of new energy sources. During the last 70 years, solar power has emerged as one of the most promising sources of renewable energy, and research has increased the power conversion efficiencies (PCE) of silicon solar cells from under 1% to over 25%. The high energetic and material costs for the required pure silicon have led to an increase in the search for viable alternative materials. The most promising material to date is the perovskite solar cell (PSC). Here the excitation takes place in the active layer (perovskite), and charge transport materials are in need to allow easy charge separation and prevent recombination. While the electron transport layers (ETLs) have been extensively studied and optimized, the hole transport layers (HTLs) emerged as the bottleneck for higher efficiencies. The synthesis and purification of the most-commonly used small molecule HTL spiro-OMeTAD are tedious and energy-intensive, and high PCEs can only be achieved by additional doping. In this thesis, various organic hole transport materials (HTMs) based on different core structures were synthesized as possible substitutes for spiro-OMeTAD. Apart from the core building block variation, different triarylamine-based donor moieties, and thiophene-based π-bridges (elongation of the conjugated system) were added to the cores to determine their influence on the optical and electrochemical properties. As a first building block, the [2.2]paracyclophane (PCP) was chosen, and eight different pseudo-*para*-substituted HTMs could be synthesized *via* CH activation. The highest occupied molecular orbital (HOMO) energy levels were determined optically, electrochemically, and theoretically, and for three of the materials, devices could be manufactured.

While all three materials displayed some extraction barriers and the implementation of dopants is still on the way, pristine Di-*p*-OMeTPAThio-PCP (**TM_31**) achieved a PCE of 8.4%. Secondly, the number of substituents was increased from two to four, and five different bis(pseudo-*meta*)-*para*-substituted [2.2]paracyclophanes could be successfully synthesized *via* NEGISHI cross-coupling. Tetra-*p*-OMeTPAThio-PCP (**TM_39**) provided the most promising results and reached 11.0% PCE as pristine HTM. Thirdly, *meso*-tetrakis-substituted porphyrins were introduced, and 17 different target materials could be synthesized with variations in donor-π-moieties and coordinated metals. As the last building block, tetraphenylethene (TPE) was introduced and directly coupled to different methoxy-substituted donor groups to determine the influence of the substitution pattern on the overall properties. The comparison of the different properties led to several different trends, with triphenylamines, methoxy groups, and 3,4-ethylenedioxythiophene (EDOT) leading to shallower HOMO energy levels compared to carbazoles, no substituents, and thiophene, respectively. Methoxy groups, *tert*-butyl groups, and more substituents increased the solubility, and by variation of the coordinated metal inside the porphyrin, the band gap and the lowest unoccupied molecular orbital (LUMO) energy level could be tuned. All in all, 35 different potential HTMs were synthesized with a HOMO energy range from -4.98 eV to -5.57 eV, thus in the required range for application in perovskite solar cells.

INHALTLICHE ZUSAMMENFASSUNG

Durch den Klimawandel und die Endlichkeit fossiler Rohstoffe ist die Entwicklung neuer Energiequellen unabdingbar. In den letzten 70 Jahren hat sich die Solarcnergic als cinc dcr viclvcrsprcchcndstcn crncucrbarcn Energien etabliert und durch extensive Forschung konnten die Wirkungsgrade von Siliziumsolarzellen von 1% auf über 25% erhöht werden. Die hohen Energie- und Materialkosten zur Herstellung von reinem Silizium haben jedoch für eine verstärkte Forschung nach Alternativen gesorgt. Der vielversprechendste Aufbau ist hier die Perowskit Solarzelle. Die Anregung findet hier in der Perowskit Schicht statt und Ladungstransportschichten werden für eine einfache Trennung der Ladungen und dem Verhindern von Rekombinationen benötigt. Während Elektronentransportschichten schon weit optimiert sind, bilden die Lochtransportschichten den aktuellen Flaschenhals für höhere Wirkungsgrade. Beim aktuellen Standard spiro-OMeTAD sind Herstellung und Aufarbeitung aufwendig und hohe Wirkungsgrade können nur durch zusätzliches Dotieren erreicht werden. Im Rahmen dieser Arbeit wurden verschiedene zentrale Bausteine, Substituenten und π-Brücken miteinander verglichen und eine Bibliothek an neuen Lochtransportmaterialen synthetisiert. Als erstes konnten acht verschiedene pseudo-*para*-substituierte [2.2]Paracyclophane synthetisiert werden. Durch optische, elektrochemische und theoretische Untersuchungen konnte die Lage der HOMO Energieniveaus bestimmt werden und drei der acht Materialien konnten in Solarzellen eingebaut werden. Während alle drei Materialien Extraktionsbarrieren offenbarten, konnte ein Wirkungsgrad von 8,4% für **TM_31** erreicht werden. Durch die Erhöhung der Substituenten wurden die Löslichkeit erhöht und die HOMO Energieniveaus leicht angehoben. Fünf verschiedene bis(pseudo-*meta*)-*para*-substituierte [2.2]Paracyclophane konnten erfolgreich hergestellt werden, für das Analogon zu **TM_31** (**TM_39**) konnte ein Wirkungsgrad von 11,0% gemessen werden. Experimente zur Dotierung sind aktuell in Arbeit. Als dritter Baustein wurden *meso*-tetrakissubstituierte Porphyrine verwendet, da diese neben der Variation an den Seitenketten auch bezüglich des koordinierten Metalls modifiziert werden können. Insgesamt 17 verschiedene Porphyrine wurde erfolgreich hergestellt. Als letzten Baustein wurde Tetraphenylethen verwendet und direkt mit verschieden substituierten Methoxy-Triphenylaminen gekuppelt. Hierbei sollte der Einfluss des Substitutionsmusters auf die Eigenschaften des Gesamtmoleküls untersucht werden. Zusammenfassend konnten verschiedene Trends bei der Variation der Seitengruppen und Metalle beobachtet werden. Die Verwendung von Triphenylaminen, Methoxysubstitution oder von EDOT sorgen für flachere HOMO Energieniveaus im Vergleich zu der Verwendung von Carbazolen, keinen Substituenten oder Thiophenen. Der Einbau von Methoxy- und *tert*-Butylgruppen erhöhte die Löslichkeit, während der Einfluss der Metalle in den Porphyrinen vor allem auf das LUMO Energieniveau und die Bandlücke zu be-obachten war. Zusammenfassend konnten 35 verschiedene Lochtransport-materialien mit HOMO Energieniveaus zwischen -4,98 eV und -5,57 eV herge-stellt werden und damit im benötigten Bereich für die Anwendung in Perowskit Solarzellen.

1 INTRODUCTION

Energy has been vital for human development and survival since the domestication of fire. Starting with the use of wood and later coal as a primary resource for heating, the industrial revolution, with the invention of steam and combustion engines, opened up a new array of applications. Naturally, the need for more energy sources increased, and the era of coal, oil, and gas began. Since then, the production and consumption of these so-called fossil energies have grown yearly except for major events such as the financial crisis in 2008 and the Covid pandemic in 2020 (Figure 1).[1]

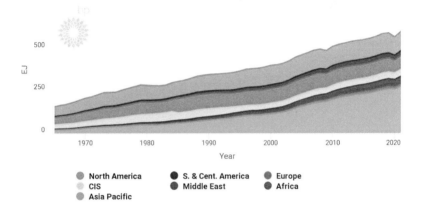

Figure 1: Energy consumption in exajoules (EJ) from 1965 to 2021. Reproduced from BP.[1]

With the increased consumption of carbon-based resources, the released amount of CO_2 and other greenhouse gases has increased to a level where climate change's effects are tangible. Having this in mind and regarding the finiteness of fossil fuels, finding alternative energy sources is of utmost importance. The shift from fossil power to renewable power sources such as hydro, wind and solar is one of the most promising and viable opportunities. Here, hydropower and wind power have faced many social pushbacks, showing that solar power, especially photovoltaics, is an attractive and decentralized option. In 2019, the world's energy demand added up to an astonishing 160 PWh, of which over 80% was

5

still generated from fossil resources.[2] With this number in mind, the global solar radiation, when only counting the landmasses, is significantly higher at 230 EWh. Harvesting only a part of the freely available energy would be sufficient to support a modern renewable energy-based industry and society. Since the late 1990s, there has been an increased effort to implement renewable energy sources into the resource grid (Figure 2), reaching up to 13% of the overall production of energy for 2021.[1]

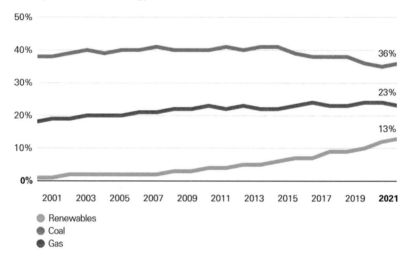

Figure 2: Global share of energy production. Reproduced from BP.[1]

Starting with the oil crisis of the 1970s, the adaption of solar energy from research purposes to commercial use began, and with the increased usage of renewable energy sources, the drive to develop more efficient and less costly devices and methods has amplified. The number of publications containing the words "solar energy" either in the title or abstract has increased yearly over the past 25 years, reaching over 26 000 publications in 2021 (Figure 3).[3]

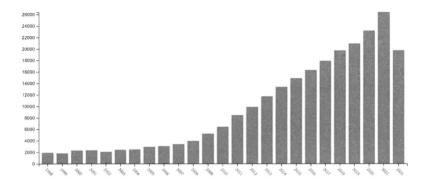

Figure 3: Results on Web of Science for "solar energy" (01.11.2022).[3]

The surge in interest in solar energy has led to enormous efforts in research to develop and optimize new materials and devices. The following chapters are meant to give an overview of the functionality of solar cells in general and to compare the different generations and device architectures, and the implementation of new materials.

1.1 THE FUNCTIONALITY OF SOLAR CELLS

The ability to harvest energy or electricity from sunlight is based on the so-called photovoltaic effect. It was first described by A. E. BECQUEREL, who observed the formation of electricity when metal electrodes were submerged into a conductive solution.[4]

Solid materials can be divided into three types of energy band structures (Figure 4a). In metals (conductors), the potential valence band (VB, blue) and conduction band (CB, green) are combined and allow a free flow of electrons. For insulators, the band gap between the valence and conduction bands is too large to be crossed even at high temperatures.[5] For semiconductors, the two energy bands (VB and CB) are separated by a small band gap, but the energy difference between the two bands can be easily crossed by external energy input, such as thermal or photo energy. Without this additional energy input, semiconductors act as insulators, and the electrons are confined to the material's crystal lattice. When the wavelength of introduced solar radiation provides enough energy to the electron to overcome the band gap, it crosses to the CB,

7

and inside the crystal lattice, a hole (positive charge) is formed. After the excitation of an electron, it thermally relaxes to the band gap and can recombine with the hole. There are three main processes for recombination: radiative recombination, recombination through defects, and Auger recombination. Radiative recombination is most common for semiconductors, where the excited electron emits a photon with a higher wavelength than the absorbed photon.[6-7] When defects are added into the semiconductor either unintentionally or intentionally (dopants), the excited electron can first relax to the energy level of the defects inside the band gap and then either recombine with the hole or get excited into the CB again.[8-9] In the third case, the Auger recombination, the relaxing electron does not emit a photon but passes the energy to a second excited electron, which is then excited higher into the CB.[10]

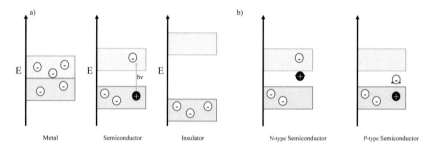

Figure 4: a) Energy band structure differences between metals, semiconductors, and insulators. b) Influence of dopants on the energy levels.

Pure semiconductors generally have an equal number of holes and excited electrons (intrinsic semiconductors) at a specific temperature and irradiation. To prevent the recombination of the excited (conductive) electrons with the holes, dopants are added (Figure 4b). As beforementioned, dopants can lead to easier recombination when the defect energy level is in the center of the band gap, but when dopants are chosen with energy levels close to either the VB or the CB, the concentration of free charge carriers can be increased, and the rate of recombination is decreased.[11] The addition of electron-rich dopants is called n-type (negative) doping, and the addition of electron-poor dopants is called p-type (positive) doping. When an active material is doped with an n-type dopant and a p-type dopant, this additionally leads to a formation of a p-n junction and charge gradients to prevent recombination by spatially separating the excited electron and the hole.

The typical process inside a solar cell, compared to an intrinsic semiconductor, is shown in Figure 5. Sunlight (4) excites electrons at the p-n junction leading to charge separation. Due to the doping of the material, the excited electron now travels through the n–doped layer (1) toward the electrode while the hole moves through the p–doped layer (3). This process prohibits fast recombination at the location of excitation. To recombine, the excited electron has to move through the attached circuit (5), allowing electricity usage.[5]

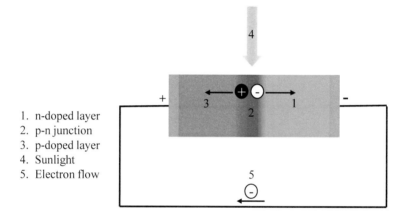

1. n-doped layer
2. p-n junction
3. p-doped layer
4. Sunlight
5. Electron flow

Figure 5: Typical cross-section of a generic solar cell.

Several parameters are needed to determine a solar cell's power conversion efficiency (η, PCE) and allow the comparison between different types, materials, and architectures. First of all, the general efficiency of a solar cell is measured as the ratio between the power that was applied (P_{in}, by irradiation) and the power that was obtained (P_{out}) (Equation 1).

$$\eta = \frac{P_{out}}{P_{in}} = \frac{FF \cdot V_{OC} \cdot I_{SC}}{P_{in}} \tag{1}$$

Equation 1: Power conversion efficiency of a solar cell.

Without quantifying the applied power, it is difficult to compare the efficiencies of two solar cells and determine whether this solar cell would benefit in ambient conditions. To allow this comparison, standard conditions have been set up as a

9

temperature of 25 °C, a light spectrum of AM1.5G, and an irradiation power of 1000 W m^{-2}. The light spectrum of AM1.5G is equivalent to the typical yearly solar spectrum of the USA on a clear day. The parameters can then be extracted from the current-voltage curve (I-V curve) by installing different resistance loads (Figure 6). With a resistance of 0 (U = 0 V), the device's highest possible I_{SC} (short circuit current) can be measured. With an infinite resistance (I = 0 A), the highest possible V_{OC} (open circuit voltage) can be determined. Additionally, the maximum power point (MPP, from $P = I * U$) is chosen as a reference. From the MPP, it is possible to calculate the current (I_{MPP}) and voltage (V_{MPP}) at which the highest power is obtained.

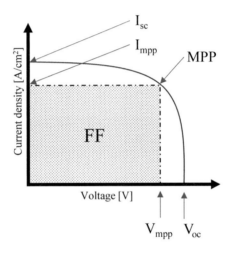

Figure 6: Typical I-V curve of a solar cell.

The ratio of the values at the MPP to the highest possible values for current and voltage is known as the Fill Factor (FF, Equation 2). The higher the Fill Factor, the higher the quality of the solar cell, as there are fewer losses in the device architecture.[11]

$$FF = \frac{MPP}{V_{OC}I_{SC}} = \frac{V_{MPP}I_{MPP}}{V_{OC}I_{SC}} \qquad (2)$$

Equation 2: Calculation of the Fill Factor (FF).

1.2 HISTORY OF SOLAR CELLS – THE FIRST GENERATION

After the first observation of the photovoltaic effect by A. E. BECQUEREL with platinum submerged into acids in 1839, this phenomenon was also observed in selenium in the 1870s and at the junction between copper and copper oxide in 1904.[4, 12-14] The first theoretical explanations of these photoelectrical effects were published by H. HERTZ in 1887 and A. EINSTEIN in 1917, with F. BLOCH developing the band theory in 1929. [15-17] After these initial steps into photo-voltaics, it took 25 more years to achieve the first commercial solar cell developed by BELL LABORATORIES in 1954.[18] Here monocrystalline silicon wafers were used as a semiconductor with gallium and lithium as dopants with an efficiency of around 6%. The following two decades, with the space race in the 1960s and the oil crisis in the 1970s, led to massive investments in photo-voltaics research.[19] In the 1990s, monocrystalline silicon solar cells reached efficiencies of over 20% using boron as a typical p–type dopant and phosphorous as a typical n–type dopant (Figure 7).[20-21]

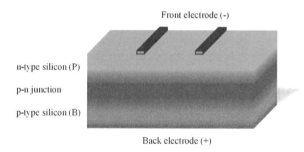

Figure 7: Cross section of a single crystal silicon solar cell.

However, monocrystalline silicon solar cells have several drawbacks. As an indirect band gap material, the silicon wafer needs a thickness of approximately 100 μm to absorb enough solar radiation, and it has to be of the highest purity. The thickness of the wafers induces increased recombination processes, and the production of these single crystals is extremely energy, material, and cost-intensive. With a band gap of 1.1 eV, single junction silicon solar cells' maximum efficiency (SHOCKLEY-QUEISSER limit) lies at 30%.[22-23] Yet, achieving this efficiency is challenging as faults in the devices, such as reflection of sunlight and blockage by the front electrodes, hinder commercial silicon solar cells from reaching a PCE over 25%. A less energy-intensive way of production is using polycrystalline silicon solar cells at the cost of efficiency.[21]

1.3   THE SECOND GENERATION – THIN-FILM SOLAR CELLS

Research continued to counter the high materials costs, and the first thin-film solar cells were developed in the 1970s by combining elements of the 13th and the 15th group of the periodic table and thus generating a p-n heterojunction (III-V heterojunction) at the interface.[24-25] Some of the most common combinations are cadmium telluride (CdTe), gallium arsenide (GaAs), or copper-indium-gallium-diselenide (CIGS) solar cells. These novel direct band gap hetero-junction solar cells only need a layer thickness of 1 to 2 μm for efficiencies of over 25%.[26-27] Stacking thin-layer solar cells into so-called multijunction solar cells can increase efficiency to over 40%, possibly reaching over 50% under extreme conditions.[28] The main issues of the III-V thin-film solar cells are the toxicity of some compounds (cadmium, arsenic) or that the compounds are scarce and therefore expensive (indium, telluride), leading to application mostly in space engineering. Another use of thin-film solar cells was the development of amorphous silicon solar cells with low efficiencies but a high enough PCE to power small handheld devices such as calculators.[29]

1.4   THE THIRD GENERATION – EMERGING SOLAR CELLS

1.4.1 Organic Solar Cells (OSCs)

Apart from developing solar cells based on inorganic semiconductors, researchers started investigating the electronic properties of organic molecules in the 1950s leading to the publication of the first semiconducting organic polymers (polyacetylene) in 1977.[30-34] Bearing in mind the same basic principle from the inorganic solar cells of generating a p-n junction, TANG developed an organic donor-acceptor solar cell in 1986 with an efficiency of 1% (Figure 8a).[35] In this case, an indium tin oxide (ITO) coated glass substrate was used as a transparent base, on which copper phthalocyanine (CuPc, donor) and a perylene tetracarboxylic derivative (acceptor) were added *via* vacuum evaporation. Soon, semiconductors were blended to increase the contact area of the p-n junction to form a bulk heterojunction (Figure 8b). This process led to an increase in efficiency to nearly 20% in 2021.[36-39]

Figure 8c demonstrates how a single heterojunction OSC works. When irradiated, electrons of the donor material are excited from the highest occupied molecular orbital (HOMO) to the lowest unoccupied molecular orbital (LUMO), similar to the process in inorganic solids with VB and CB. To prevent recombination, the acceptor is chosen according to the energy level of the LUMO. Now the excited electron can travel from the LUMO of the donor through the LUMO of the acceptor to the electrode *via* charge transfer. The same process is applied for the hole traveling to the ITO.

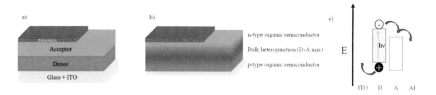

Figure 8: a) Cross section of a two-layered organic solar cell. b) Cross section of a bulk heterojunction organic solar cell. c) Distribution of HOMO and LUMO levels in OSCs.

Some major advantages of organic solar cells compared to silicon solar cells are their possible cheap production through low-temperature methods, high versatility *via* organic chemistry, low material costs, and applicability in new products such as flexible thin films.[40-41]

1.4.2 Dye-Sensitized Solar Cells (DSSCs)

Another charge generation and separation approach has been implemented in dye-sensitized solar cells (DSSCs). Here, a dye is used as active material and sandwiched between an electron-conducting layer and a hole-conducting layer. Therefore the junction is no longer p-n type but p-i-n type. B. O'REGAN and M. GRÄTZEL first described this device architecture in 1991 with a PCE of around 7%.[47] Typically mesoporous titanium dioxide (TiO_2) is used as an electron-conducting layer, and the dye is introduced into the pores of the TiO_2. Additionally, a liquid electrolyte is used as a hole-conducting layer (Figure 9a). When using redox-active electrolytes, PCEs of over 10% can be achieved.[43-45] As with the OSC's energy levels, it is important to use adequate electrolytes and dyes to prevent recombination and losses in PCE (Figure 9b).

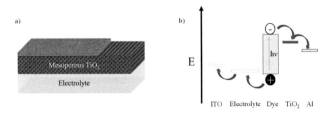

Figure 9: a) Cross section of a dye-sensitized solar cell (DSSC). b) Energy levels of the components of a DSSC.

While DSSCs are easily processable and the dyes possess high photon absorbance in specific wavelengths, they are not applicable for widespread use as the implementation of liquid electrolytes offers issues with freezing at low temperatures and expansion at high temperatures. In addition, the dyes do absorb efficiently only in specific wavelength areas. To achieve high PCEs, the absorbed spectrum needs to be broader.

1.4.3 Perovskite Solar Cells (PSCs)

The most promising type of emerging solar cell is the perovskite solar cell. It could be viewed as the solid-state counterpart of the DSSC. It combines some of the advantages of the different beforementioned types of solar cells: a solid hole transport material as in the organic solar cells, a perovskite with a broad absorbance of visible light, and metallic oxides (mostly TiO_2) as electron transport material as in dye-sensitized solar cells (Figure 10a).[46] Depending on the stability and solubility of the charge transport materials, the general architecture of the solar cell can vary. As mentioned for the DSSCs already, the conventional build-up is called n-i-p, with the electron-transport layer (ETL, n-type) on top of the ITO, then the active layer (i) and then the hole-transport layer (HTL, p-type) between the active layer and the Ag/Au electrode. For transparent HTMs, this architecture can be inverted (p-i-n). Typically, perovskites of the general structure ABX_3 are applied, such as methyl ammonium (MA, $CH_3NH_3^+$) or formamidinium (FA, $CH(NH_2)_2^+$) for position A, lead (Pb) or tin (Sn) for position B and halogenides such as iodide, bromide or chloride for position X.[47] Typically, the energy levels of a $MAPbI_3$ (methyl ammonium lead iodide) solar cell range from around -5.5 eV for the VB to -3.95 eV for the CB.[48]

Figure 10: a) Cross section of a n-i-p perovskite solar cell (PSC). b) Ideal cubic perovskite structure ABX₃. c) Schematic view of the energy levels of the components of a perovskite solar cell.

The first ever published perovskite solar cell by KOJIMA *et al.* from 2009 still used a liquid electrolyte and reached a PCE of 3.8%, which increased to 6.5% by 2011.[49-50] The breakthrough followed within the next years by replacing the liquid electrolyte with solid hole transport materials, leading to PCEs of 10% in 2012.[51-52] Current high-efficiency perovskite solar cells are already in the range of 25%, showing incredible growth in only 13 years.[53] Still, some huge issues cannot be denied. The volatility of its organic compounds causes the perovskite to be an unstable layer in the cell. Additionally, the more efficient perovskite solar cells use the potentially toxic compound lead as part of the crystal structure. Some research efforts try replacing lead with more environmentally friendly and less toxic metals.[54-55] Apart from the choice of perovskite structure, the largest levers in increasing the potential PCE are in varying electron transport layer and hole transport layer.

Another novel approach to increase the PCE in silicon and perovskite solar cells is the combination of both cells into a tandem solar cell. This is possible because silicon and perovskite have different absorption maxima, and the active silicon layer can harvest the radiation that passes through the perovskite. With this, PCEs of approximately 30% are reported, and by optimizing the junctions, PCEs of up to 40% seem to be within reach.[56-59]

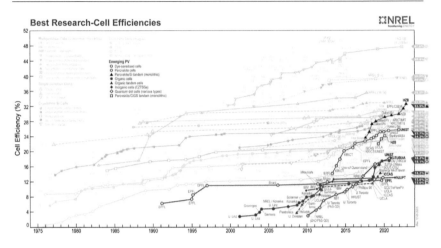

Figure 11: Overview of verified PCEs with highlighted results for emerging solar cells. Reproduced from NREL.[60]

The sparked interest in perovskite and perovskite tandem solar cells can be visualized best by the yearly updated Best Research-Cell Efficiencies overview of the NATIONAL RENEWABLE ENERGY LABORATORY showing the steep and incomparable increase in PCEs of PSCs over the last 10 years (Figure 11, yellow-filled circle for PSCs and blue-filled triangles for perovskite tandem solar cells).

1.5 CHARGE TRANSPORT MATERIALS – HOLE TRANSPORT MATERIALS (HTMs)

Charge transport materials are needed to allow a simple transfer of the generated electrons and holes from the active layer of the solar cell to the corresponding electrodes. By implementing charge transport materials, recombination at the interfaces is reduced and ideally prevented. This chapter will briefly summarize electron transport materials and inorganic hole transport materials while focusing on organic hole transport materials, especially small molecules.

1.5.1 Electron Transport Materials (ETMs)

The electron transport materials are typically metal oxides, with titanium(IV) dioxide (TiO_2) being the most commonly used material.[61] This is due to their extensive investigation during the optimization of DSSCs. The CB energy of TiO_2 with -4.1 eV is close to the CB energy of the perovskites while slightly lower to allow easy electron injection.[62] With a band gap of around 3.0 eV, depending on the crystal structure of the TiO_2, the VB energy is deep enough to prevent accidental hole extraction.[63-65] However, the electron mobility is limited in pure TiO_2 films, and different doping strategies are applied to achieve higher electron mobilities and, therefore, higher PCEs.[66] Some examples of the doping process are rubidium chloride, niobium, or chlorine.[67-69] Additionally, TiO_2 needs high processing temperatures, eliminating them as ETM for solution-processed PSCs.[70] Another promising ETM is tin(IV) di-oxide (SnO_2), which profits from a similar conduction band energy but has a lower processing temperature and a higher electron conductivity than TiO_2.[71-73] Similar to TiO_2, the electron conductivity can be increased by adding dopants, such as rare-earth metals or ethylene diamine tetra-acetic acid (EDTA).[74-75] Further often applied inorganic ETMs are zinc oxide (ZnO), barium stannate ($BaSnO_3$), or metal sulfides.[76-78] Apart from inorganic ETMs, organic ETMs can also be implemented in the device, mostly based on fullerene and its derivatives, due to suitable energy levels and low fabrication temperatures.[79] Some typical examples are C_{60} or $PC_{61}BM$ ([6,6]-phenyl-butyric acid methyl ester).[80-82]

1.5.2 Inorganic Hole Transport Materials

Inorganic hole transport materials are getting increased attention in research due to their higher thermal stability than organic HTMs, but they still lack efficient hole conduction to replace organic HTMs. The PCEs have increased from under 1% in 2013 to approximately 20%.[83-84] Some of the widely applied materials are nickel oxides (NiO_x), copper compounds (CuX), and molybdenum oxides (MoO_x). For these materials, the energy level of the VB is important, while the band gap only has to be large enough to prevent electron extraction; for NiO_x, for example, the VB energy is between -5.2 eV and -5.4 eV.[85] For these in-organic HTMs, the deposition methods play a huge role in the overall PCE, as the interface connection with the perovskite depends on the morphology of the HTM. In contrast to inorganic ETMs, the HTMs can be processed *via* solution. For NiO_x, the solution processability increased PCE from 9.5% as a mesoporous thin film to 16.0% as nanoparticles and 18.0% as highly transparent thin films.[86-88] To further increase the PCE, inorganic HTMs can be doped with cations such as Cu^{2+} (PCE 19.8%), Li^+ (PCE 19.9%), Co^{2+} (PCE 18.6%), Cs^+ (PCE 19.3%) or Zn^{2+} (PCE 19.6%).[89-93] Copper compounds can also be easily processed *via* solution processes, and depending on the exact composition, these HTMs can achieve PCEs of up to 19%.[94] Some examples are Cu_2O (PCE 13.4%), CuO_x (PCE 19.0%), CuS (PCE 16.2%), and $CuSCN$ (PCE 18.8%).[95-98] Another interesting compound as an inorganic HTM is MoO_x, which is stable, cheap, and non-toxic. Although using only MoO_x does not result in high PCEs (13.1%), the material shows strong hole extraction and can be used as an interface between organic HTM and electrode.[99-101]

1.5.3 Organic Hole Transport Materials

While inorganic hole transport materials profit from high stabilities and show increases in performance, much effort is still needed to reach the PCEs from perovskite solar cells built with organic hole transport materials. These organic hole transport materials can be divided into two main groups, the polymers, and the small molecules. Depending on the properties of these HTMs, the architecture of the perovskite solar cells can either be conventional (n-i-p) or inverted (p-i-n). For most small molecules with high solubilities in the solvents that are used to add the perovskite, the conventional structure is preferred, while for polymers with large enough band gaps to not absorb in the same range of wavelength as the perovskite, the inverted structure is preferred.

1.5.3.1 *Polymers as Hole Transport Materials*

The use of polymers as acceptors has been widely implemented during the research on organic solar cells. The first time an organic polymer was implemented into a perovskite solar cell was by JENG *et al.* in 2013, who used poly(3,4-ethylenedioxythiophene):poly(styrenesulfonate) (PEDOT:PSS, Figure 12a).[81] PEDOT:PSS provides a wide band gap with the HOMO energy level at -4.9 eV and the LUMO energy level at -2.2 eV, thus not absorbing light in the visible spectrum and allowing an inverted architecture.[102] However, pristine PEDOT:PSS shows some significant disadvantages, such as a high acidity due to the sulfonates, poor hole mobility, and a difference of 0.5 eV compared to $MAPbI_3$ with a VB energy level of -5.4 eV.[103] The addition of dopants could weaken the impact of both issues. ZOU *et al.* developed a doping strategy in which the sodium salt of PSS (PSSNa) was added as an electrolyte to adjust the energy levels and increase the hole mobility of the polymers (Figure 12b).[104] The group could show that depending on the ratio of PSSNa, the HOMO energy level of PEDOT:PSS shifted from -4.9 eV up to -5.2 eV and, therefore, closer to the VB energy level of the perovskite, increasing the hole extraction and the PCE from 12.3% to 15.5%. Another technique involved the addition of 2,3,5,6-tetrafluoro-7,7,8,8-tetracyanoquindodimethane (F4-TCNQ) as dopant, which has been well researched in HTMs for organic light-emitting diodes (OLEDs).[105] The low LUMO energy level (-5.24 eV) of F4-TCNQ as an inter-layer between PEDOT:PSS and the perovskite leads to a better hole extraction and the PCE could be improved from 13.3% to 17.2% with 0.3 wt% doping (Figure 12b).[103] A similar effect could be achieved by adding so-called self-organized hole extraction layers (SOHEL). Here, a perfluorinated ionomer (tetrafluoroethylene-perfluoro-3,6-dioxa-4-methyl-7-octene-sulfonic acid co-polymer) allowed a specific fine-tuning of the HOMO energy level of the HTM with an increase of PCE from 8.1% to 11.7%.[106] Apart from adding chemical dopants, as just discussed, the stability and PCE could be improved by changing the pH level of PEDOT:PSS. [107] WANG et al. showed that with the addition of imidazole (Figure 12b), the HOMO energy level changed from -4.9 eV for the acidic PEDOT:PSS to -5.0 eV for the neutral PEDOT:PSS to -5.3 eV for alkaline PEDOT:PSS and thus increasing the PCE from 12.7% to 15.7%.[108] Similar effects were observed when treating the PEDOT:PSS layer with ammonia.[109]

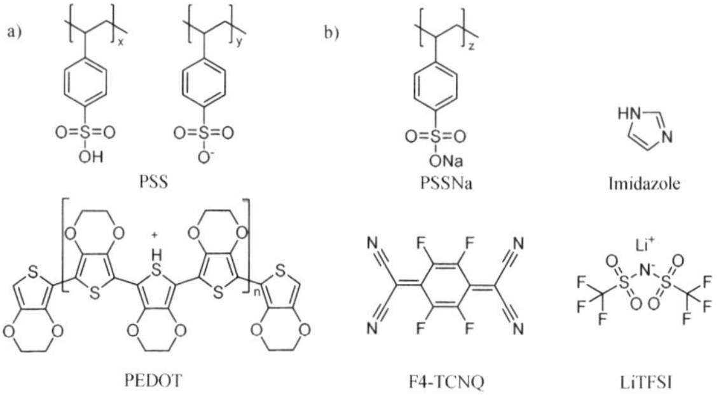

Figure 12: a) Chemical structure of PEDOT:PSS. b) Typical dopants.

With PEDOT:PSS showing the beforementioned discouraging issues and the need for doping to achieve competitive PCEs, research continued searching for different polymers with better fitting energy levels. Polytriarylamines have been known as good HTMs for OLEDs and also show better fitting energy levels compared to PEDOT:PSS with a HOMO energy level of -5.2 eV and a LUMO energy level of -1.8 eV for poly(bis(4-phenyl)(2,4,6-trimethylphenyl)amine (PTAA, Figure 13).[102, 110-111] Due to the good electronical properties and fitting energy levels, doping with F4-TCNQ can be applied, but it is not as necessary as in PEDOT:PSS.[112] For PTAA the application method and solvent use have greatly influenced the PCE by promoting the growth of defect-free perovskite crystals.[113-114] The crystal growth of the perovskites can either be enhanced by prewetting the hydrophobic surface of PTAA with *N,N*-dimethyl formamide (DMF) or by adding interlayers between the PTAA film and the perovskite, leading to PCEs of over 19%.[115-117]

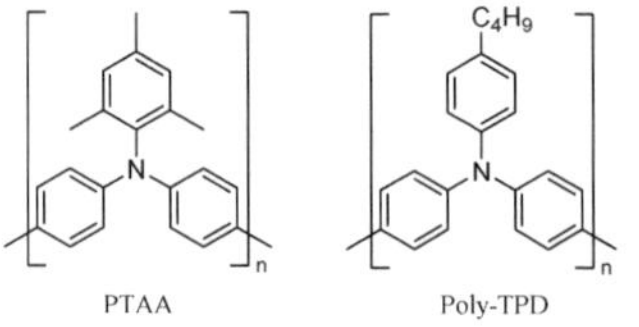

Figure 13: Triarylamine-based HTMs PTAA and poly-TPD.

Another triarylamine-based HTM is poly(*N,N'*-bis(4-butylphenyl)-*N,N'*-bis(phenyl)benzidine) (Poly-TPD, Figure 13). The HOMO energy level is similar to the perovskite (-5.4 eV), and the growth of large perovskite grains is promoted, increasing the PCE to 15.4%.[118] By optimizing the wettability, the PCE was increased to 22.1% by ZHANG *et al.* in 2021.[119-120]

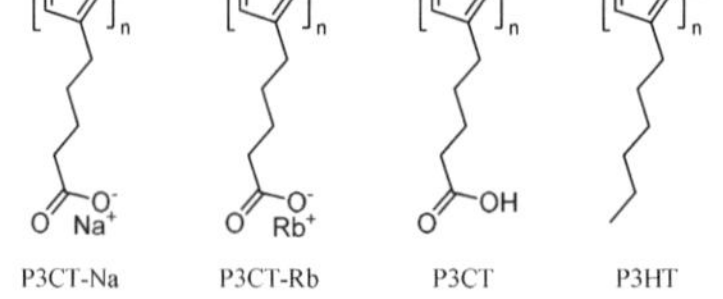

Figure 14: Different derivatives of thiophene-based polymers.

Like the beforementioned PEDOT:PSS, where poly(3,4-ethylenedioxy-thiophene) is implemented as one copolymer, pure thiophene polymers can also be applied as HTM. The most well-known thiophene polymer is poly(3-hexylthiophene) (P3HT, Figure 14). P3HT can convince with a fitting HOMO energy level of -5.20 eV and high charge mobility, allowing pristine P3HT to achieve a PCE of 10.8%.[121] With the addition of F4-TCNQ or lithium bis(trifluoromethanesulfonyl)imide (LiTFSI, Figure 12b) as dopants, the PCE could be increased to 14.4% and 17.5%, respectively.[122-123] Further modification of the interface between the perovskite and P3HT increased PCE to 23.3%.[124] Another thiophene-based hole transporting polymer is the sodium salt of poly[3-(4-carboxybutyl)thiophene-2,5-diyl] (P3CT-Na, Figure 14), with a PCE of 16.6%.[125] However, the non-alignment with the VB of the perovskite (HOMO energy level of -5.26 eV) and a strong tendency to aggregate fueled further investigations. The PCE could be improved to 19.6% by LI et al. by replacing the counterion Na^+ with $CH_3NH_3^+$, thus breaking up the aggregation.[126] This PCE was further increased to 20.5% by LI et al. when Rb^+ and Cs^+ were introduced as counterions and reduced the HOMO energy level to -5.35 eV (P3CT-Rb) and -5.32 eV (P3CT-Cs).[127] The use of pristine P3CT without any salts was first described by LIAO et al. in 2021.[128] In this case, the carboxylic groups of the side chain are anchored to the ITO electrode and the perovskite, leading to much lower recombination *via* faster charge extraction and increasing the PCE to 21.3%.

1.5.3.2 Small Molecules as Hole Transport Materials

Compared to polymers, the synthetic options for structure modification and optimization of small organic molecules increase manifold. Target molecules can be tailored to fit the required applications. For organic hole transport materials, that means high thermal stability, good solubility in solvents that do not impact the perovskites, no crystallinity, fitting HOMO energy levels (-5.4 eV to -5.0 eV) compared to the implemented perovskite ($MAPbI_3$ VB ~ -5.5 eV) and the ability to transport the desired charges. The current benchmark small molecule in PSCs is 2,2',7,7'-tetrakis(*N,N*-di-*p*-methoxyphenylamine)-9,9'-spirobifluorene (spiro-OMeTAD) with a certified PCE of 23.3%.[53] The spirobifluorene was first introduced into organic optoelectronics by SALBECK et al. in 1997 when spiro-TAD and spiro-OMeTAD were published as hole carriers in light-emitting diodes

(Figure 15).[129] Due to the difference in glass transition temperature T_g between spiro-TAD and spiro-OMeTAD (62 °C *versus* >100 °C) and solubility, spiro-OMeTAD was then used as an electrolyte in DSSCs by BACH *et al.* in 1998 where a high photon-induced current was measurable, but the overall PCE only reached 0.7%.[130] In 2012, spiro-OMeTAD was first introduced into PSCs and achieved a PCE of 9.7% without adding any dopants.[51]

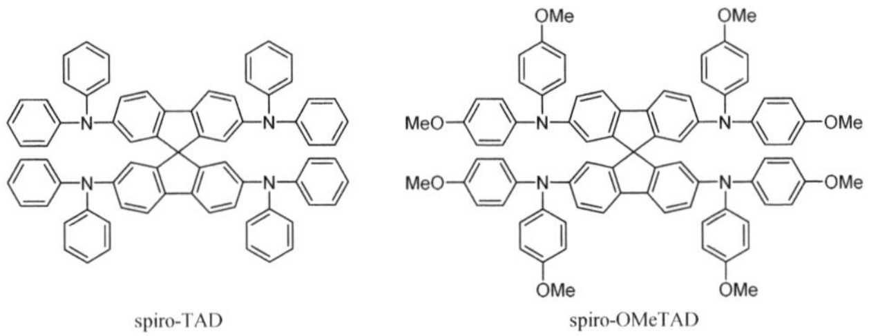

Figure 15: Structures of spiro-TAD and spiro-OMeTAD.

The rapid success of spiro-OMeTAD fueled the interest in further understanding the opportunities and possibilities of spiro-based small molecules. In 2016, HU *et al.* published three new spiro-based derivatives in which the methoxy groups were substituted with ethyl (spiro-E), *N,N*-dimethylamino (spiro-N), and methylsulfanyl (spiro-S) groups to determine changes in the energy levels (Figure 16).[131] The group could show that the HOMO energy levels changed quite drastically from -4.76 eV for spiro-OMeTAD to -4.81 eV for spiro-E, -4.42 eV for spiro-N and -4.92 eV for spiro-S. By closing the difference to the VB energy level of the perovskite, spiro-S showed a higher PCE than spiro-OMeTAD in device performance (15.9% versus 11.5%). Apart from substituting the methoxy groups, JEON *et al.* showed that positioning the methoxy group on the phenyl ring also influenced the energy levels (Figure 16).[132] By switching half of the methoxy groups from the *para* to the *ortho* position, the PCE could be increased from 14.9% for doped spiro-OMeTAD to 16.7% in doped spiro-(op)OMeTAD. This effect was further investigated by extensive calculations of MURRAY *et al.* showing that using only the *ortho* positions would decrease the HOMO energy level to -5.43 eV.[133]

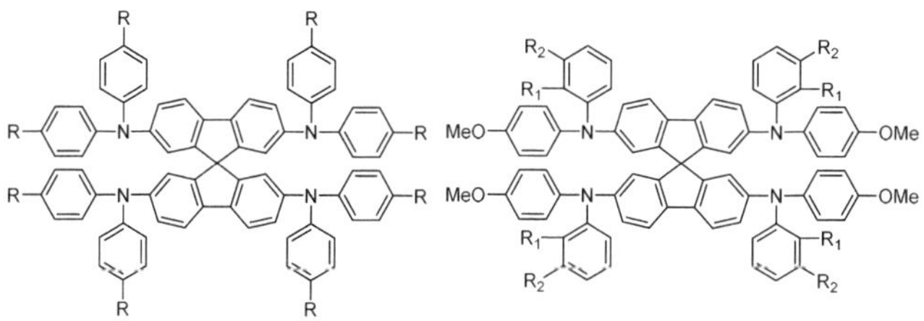

Spiro-E: R = Et
Spiro-N: R = NMe₂
Spiro-S: R = SMe

Spiro-(mp)OMeTAD: R₁ = H, R₂ = OMe
Spiro-(op)OMeTAD: R₁ = OMe, R₂ = H

Figure 16: Spiro-TAD derivatives with differences in substitution pattern and end groups.

As beforementioned, the group of JEON *et al.* reached a PCE of 16.7%, but only with doping. One of the two main disadvantages of spiro-OMeTAD is the low hole conductivity. To combat this restrictive property, chemical doping opens a broad scope of possibilities. The easiest and most used method of increasing the hole conductivity in spiro-OMeTAD was described first by SNAITH *et al.* in 2006 when the addition of ionic LiTFSI increased the hole mobility by factor 10, and this effect was later ascribed to the presence of oxygen and LiTFSI allowing the oxidation of one of the nitrogen atoms.[134-135] However, the lithium salts are hygroscopic and tend to degrade when water is in the vicinity, while the formed Li^+ can diffuse to the ETM and decrease the thin-film morphology of the HTM.[136] This effect is intensified when the working temperature increases.[137] To counter the degradation of LiTFSI with water, the most commonly used additive is *tert*-butyl-pyridine (TBP). The nitrogen of the pyridine can coordinate the lithium and prevent the reaction with water; additionally, the TBP acts as a morphology controller and prevents the agglomeration of LiTFSI in the hole transport layer.[138-141] Still, adding TBP creates a new issue, as it can dissolve PbI_2 and degrade the perovskite.[140] An interesting new approach was published by TAN *et al.* in 2019 when the group synthesized the spiro-OMeTAD(TFSI)₂ salt as a dopant and combined it with pristine spiro-OMeTAD in the device, leading to a PCE of 19.3% and not having to deal with the issues caused by the lithium ions.[142] Apart from the doping with LiTSFI, many other metals and metal complexes can be used, but none of them reach higher PCEs than LiTFSI. An overview of different doping strategies can be found in the reviews of REN *et al.* for spiro-OMeTAD in particular and SCACCABAROZZI *et al.* for more general practices in organic semiconductors.[143-144] The second main disadvantage of spiro-OMeTAD is the synthesis. Starting from a halogenated biphenyl, the 9,9'-spirobifluorene is synthesized *via* a GRIGNARD reaction, with yields not surpassing 60%. In a second step, the 9,9'-spiro-bifluorene is brominated with elemental bromine and iron(III) chloride as a catalyst in chlorinated solvents (CH_2Cl_2, $CHCl_3$, CCl_4) *via* electrophilic aromatic substitution with yields of up to 90%.[145] In the last step, the brominated 9,9'-spirobifluorene undergoes a palladium-catalyzed BUCHWALD-HARTWIG amination with the corresponding 4,4'-dimethoxydiphenylamine with yields lower than 50%.[132] Each of the steps requires expensive and toxic materials as well as extensive work-ups. Therefore further investigations into possible new hole transport materials were necessary to either develop new materials without the need for dopants or to modify the existing materials to decrease their cost of raw materials.

Some interesting new materials have been developed based on the spiro structure, which combines the two π-systems while twisting the two sides due to the sp^3 hybridization of the central carbon atom. SALIBA *et al.* published a simpler synthesis for an asymmetric spiro-connected fluorene-dithiophene (FDT, Figure 17 left).[146] By implementing sulfur into the structure, defects at the interface between HTM and perovskite were better passivated, and with only two phenylamine groups present, the yield of the palladium-catalyzed cross-coupling reached over 80%. By doping with LiTFSI, the PSC could reach a PCE of 20.2% and outperform spiro-OMeTAD in similar devices. In 2021, ABDELLAH *et al.* published another similar spiro-based HTM (Spiro-IA, Figure 17 right), which could reach a PCE of 15.6% without any dopants.[147] By employing a spiro-xanthene-fluorene core instead of the beforementioned bifluorene, the GRIGNARD reaction could be skipped, and the further steps were achieved *via* MIYAURA borylations and SUZUKI-MIYAURA couplings under more gentle conditions.

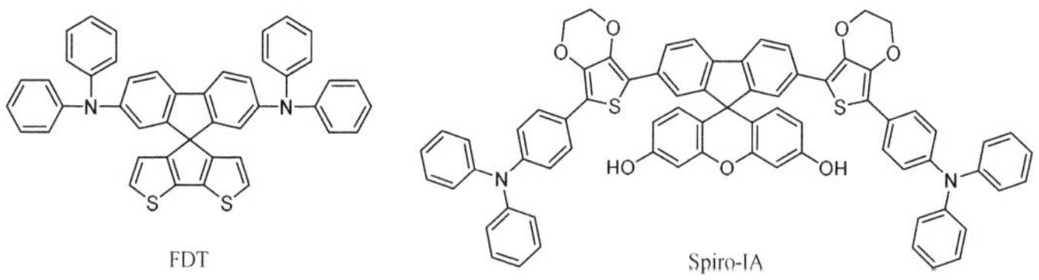

Figure 17: Spiro-based HTMs FDT (left) and spiro-IA (right).

Besides the spiro-based structures, several core building blocks have been employed in PSCs. Some of these building blocks are fused aromatic systems such as anthracenes, pentacenes, and pyrenes. The planarity of this central building unit allows π-π stacking and well-structured thin films. To increase solubility and hole mobility, side groups can be attached to the central building block. In 2015, KAZIM *et al.* published pentacene with triisopropylsilyl groups (TIPS) and achieved a PCE of 11.8% without dopants (Figure 18 left).[148] The group of PHAM *et al.* published three acene-based structures containing methoxy-substituted triphenylamines (TPA). For TPA-NADT-TPA they could achieve PCEs of up to 10.2% in inverted architecture (p-i-n) PSCs (Figure 18 right).[149]

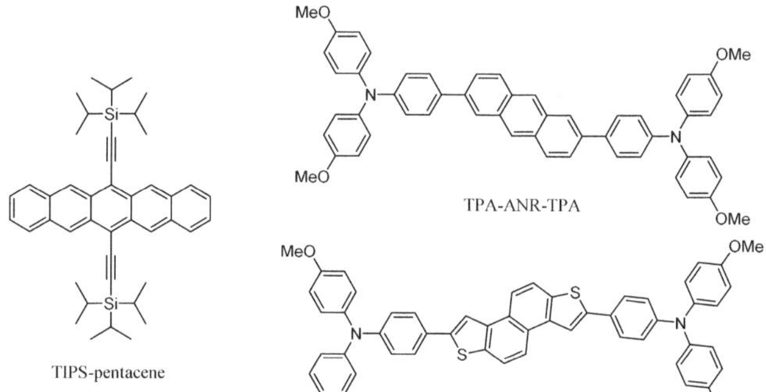

Figure 18: Acene-based organic HTMs.

GE *et al.* and QIU *et al.* continued on the same basic principle but employed pyrene instead of linear acenes. GE *et al.* synthesized a tetra-substituted pyrene with thiophene-TPA as side groups (OMe-TATpyr, Figure 19 left). These side groups increased electron density and solubility compared to only using dimethoxydiphenylamines as donor groups and increased the PCE from 8.8% to 20.0%.[150] A similar approach was taken by QIU *et al.* in which they used di(dimethoxytriphenylamine)amine as electron-rich groups (PYR16 and PYR27). Additionally, the possible rotation of the triarylamines increased the solubility.[151] They investigated the positioning of the donor groups on the pyrene moiety and proved that substitution on the 1 and 6 positions of the pyrene (PYR16, Figure 19 right) shows higher PCEs (17.0% versus 14.6%) and stability than on the 2 and 7 positions (PYR27).

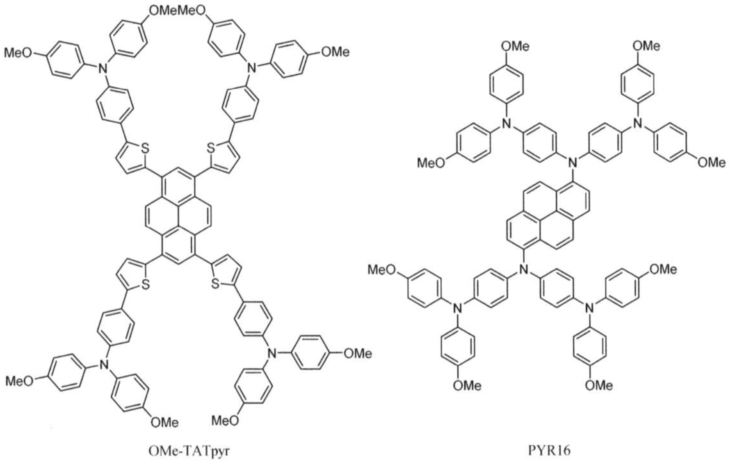

Figure 19: Pyrene-based organic HTMs.

In addition to fused aromatic systems, triphenylamine is another well-established building block that has already shown its use as an electron-donating group in many of the beforementioned examples. The triphenylamine can also be implemented as a core in three-dimensional star-shaped molecules. Furthermore, the triphenylamine can easily be modified on all its aromatic ring systems. For the easiest reaction conditions, tris(4-bromophenyl) amine is used as a starting material and can then be modified at all three bromo positions *via* the typical and well-known cross-coupling reactions. CHANG *et al.* synthesized four star-shaped hole transport materials, the best-performing HTM is shown in Figure 20 (YC-01).[152] By taking advantage of the acidic proton on EDOT, they got satisfactory yields *via* palladium-catalyzed CH activation compared to the literature procedure employing a STILLE coupling and not only worked without stannanes and reduced the reaction from 6 steps to only 2 steps.[153] The HOMO energy level of -5.28 eV is close enough to the used perovskite (VB energy -5.43 eV), and the group could show PCEs of up to 17.5% and therefore similar to the doped spiro-OMeTAD (PCE 17.8%). Four years earlier, in 2014, CHOI *et al.* could show that the extension of the π-system of the HTM lowers the HOMO energy level and allows better fine-tuning of the needed optoelectronic properties.[154] They also started from tris(4-bromophenyl) amine but then used a diphenylethyl borolane as a coupling partner in a SUZUKI-MIYAURA coupling. The final product (TPA-MeOPh, Figure 20) showed a HOMO energy level of -5.29 eV, fitting very well to the VB energy of the perovskite and leading to a PCE of 7.4% without dopants and 10.8% with LiTFSI and TBP doping. In 2018 LIU *et al.* published interesting results on another star-shaped triarylamine (*m*-MTDATA, Figure 20), which due to its exceptional solubility and hydrophobic character, could be employed as HTM in p-i-n device architectures.[155] The devices built with pristine *m*-MTDATA achieved a PCE maximum of up to 17.7%. And while some of these HTMs achieve high PCEs without the need for dopants, CHO *et al.* could show that the role of LiTFSI and TBP as dopants is comparable to spiro-OMeTAD, where one of the nitrogens of the triarylamines is oxidized by O_2 and stabilized by the TFSI anion and TBP.[156] They increased the PCE from 1.1% for the pristine HTM BT41 (Figure 20) to 9.0% for the doped hole transport layer.

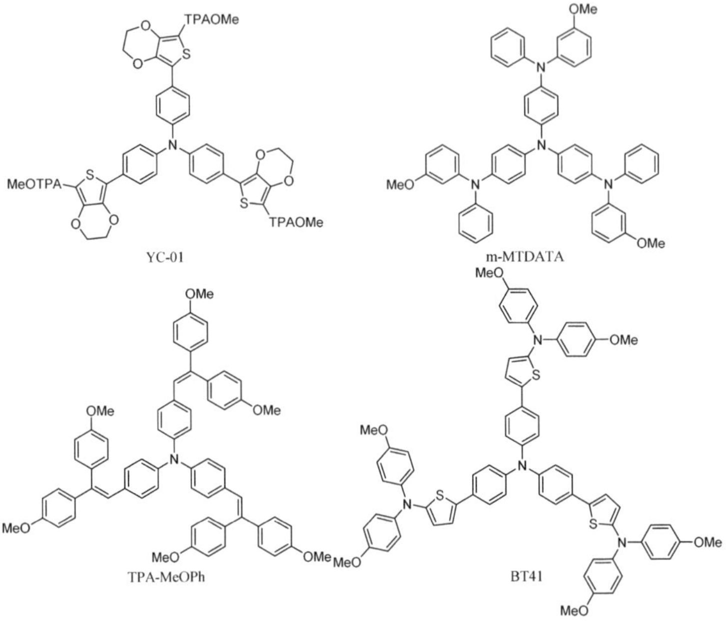

Figure 20: Triphenyl-based organic HTMs.

Another interesting building block is the [2.2]paracyclophane (PCP). It was first discovered in 1949 by BROWN and FARTHING *via* gas-phase pyrolysis of *para*-xylene, and in 1951, the first synthesis *via* intramolecular cyclization was reported by CRAM *et al.*.[157-158] The two aromatic benzene rings are connected *via* ethylene bridges and have a distance of only 3.09 Å. Even though the aromatic π-system is technically divided by the ethylene bridges, the close vicinity of the two benzene rings allows a through-space conjugation.[159] The structure of [2.2]paracyclophane allows the synthesis of different isomers with planar chirality, which made the PCP an interesting building block for thermally activated delayed fluorescence (TADF) and additionally as a source for circularly polarized luminescence (CPL) in OLEDs.[160-162] Apart from the application in OLEDs, the [2.2]paracyclophane has also been used as a core for different hole transport materials in PSCs. PARK *et al.* published the first report in 2015, where the bis(pseudo-*meta*)-para-brominated [2.2]paracyclophane was coupled with 4-pinacolboryl-*N,N*-bis(4-methoxyphenyl)aniline in a SUZUKI-MIYAURA coupling to yield PCP-TPA (Figure 21 left).[163] The PCP core allows tightly packed thin films while the triarylamine side groups increase the solubility and hole extraction properties. The HOMO energy level of PCP-TPA was observed at -5.33 eV with a band gap of 2.96 eV and, therefore, promised a good fit for the typical perovskites. When doped with LiTFSI, PCP-TPA showed a very competitive PCE of 17.6% compared to spiro-OMeTAD (15.4%) under the same conditions. In 2016, PARK *et al.* compared the photovoltaic performances of the tri-substituted PCP (Tri-TPA) and tetra-substituted PCP-TPA with a linear di(bis(4-methoxyphenyl)aniline-substituted benzene (Di-TPA).[164] They could show that the charge transport out-of-plane was highest for PCP-TPA and that the multiple arms of the PCP increase intermolecular charge transport. In 2021, LIN *et al.* published another interesting report on the properties of PCPs as HTMs.[165] They could show that the symmetrical bis(pseudo-*meta*)-para substitution pattern on the PCP is superior to substitution in a bis(pseudo-*meta*)-ortho pattern and by adding a vinyl bridge in between the triarylamines and the PCP (WS-1, Figure 21 right), they could show an increase in PCE by increasing the π-system (19.1% versus 17.8% for PCP-TPA).

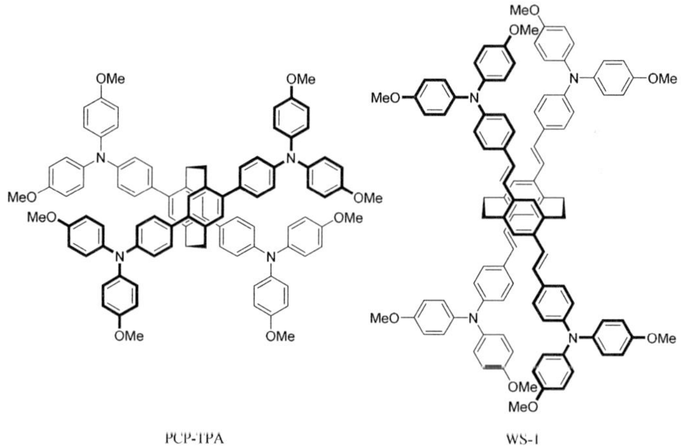

Figure 21: [2.2]paracyclophane-based organic HTMs.

A further intriguing electron-rich molecule that can be used as a core structure is tetraphenylethene (TPE). It was first implemented into perovskite solar cells by KE *et al.* in 2018 (TPE-1, Figure 22 left).[166] The TPE was substituted with four di(bis(4-methoxyphenyl)-aniline moieties and reached a PCE of 7.2% without dopants and a novel lead-free tin-based perovskite. In 2019, SHEN et al. synthesized a similar molecule but used tetrathienylethene (TTE) as the central building block and compared TTE-1 with the semi-locked TTE-2, in which they fused two thiophene moieties (Figure 22 right).[167] The increased planarity led to better morphology control and higher hole conductivity, increasing the PCE from 13.6% to 20.0%. The same effect could be described by ZHANG et al. for the TPE core in 2020, where the fusion of the benzene rings on each side and the following increased planarity increased the PCE from 15.8% to 18.3% (F-TPE, Figure 22 left).[168] The different fusing possibilities were further investigated and published by ZHANG et al. in 2021.[169]

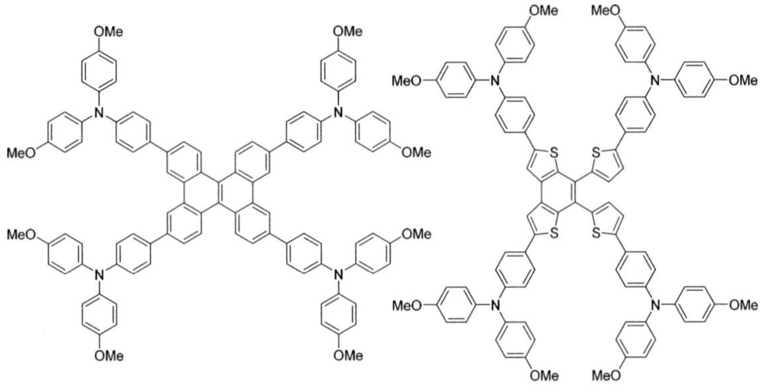

Figure 22: TPE-based organic HTMs.

The last group of molecules discussed in this introduction are porphyrins as a core structure. The porphyrin structure consists of 18 π-electrons and is built up from four pyrroles and four methine carbons. The first synthesis was described by ROTHEMUND and MENOTTI in 1941 and further optimized by ADLER et al. in 1964.[170-171] Pyrrole and benzaldehyde were refluxed in propionic acid to yield the symmetrical *meso*-benzyl-substituted porphyrin for this synthesis. In 2001, the first experiments regarding hole conductivity were performed by SAVENIJE et al. as the large planar core of the porphyrin is prone to π-π stacking and could, therefore, increase the intermolecular charge mobility.[172] HERNÁNDEZ-FERNÁNDEZ et al. showed in a theoretical study that the metals inside the porphyrins are significant for hole conductivity and that the metals that do not influence the HOMO increase conductivity (Zn, Ti, Cd, and Pd) while Ni, Fe, and Pt decrease the conductivity due to their influence on the HOMO.[173] In 2017, CHEN et al. synthesized a symmetrical meso-substituted porphyrin by using bis(4-methoxyphenyl)-aminobenzaldehyde and adding copper and zinc as coordinated metals (ZnP and CuP, Figure 23).[174] Here CuP showed a low HOMO energy level of -5.37 eV while the HOMO energy level of ZnP was higher with -5.29 eV, therefore closer to the VB energy of the perovskite and achieving a PCE of 17.8% compared to CuP with 15.3%. In the same year, LEE et al. investigated the ZnP further by adding the methoxy group free triphenyl-amine as donors and by adding pyridine into the scaffold of the aldehyde.[175] With these differences in donor structure, the range of HOMO energy levels went from -5.30 eV for PZn-DPPA to -5.08 eV for PZn-TPA-O (Figure 23). The best PCEs were achieved with PZn-DPPA (14.1%) and PZn-DPPA-O (13.5%), which was comparable to their spiro-OMeTAD standard (14.6%) and showed that the PCE is strongly dependent on how well the energy levels fit together to allow good hole extraction and conductivity. The tuning of properties was con-tinued by AZMI et al. by adding fluorine as a substituent to the benzaldehyde before condensation.[176] They could show that the position of the fluorine corresponds to that the benzene ring's electron density strongly influences the HOMO energy level, with PZn-2FTPA having a HOMO energy level of -5.40 eV while PZn-3FTPA had a HOMO energy level of -5.29 eV (Figure 23). For the PCE, PZn-3FTPA reached 17.7% while PZn-2FTPA triumphed with 18.8%. In addition to the symmetrical porphyrins, the linear derivatives were also investigated. KANG et al. increased the length of the π-system by adding an alkyne bridge between the porphyrin and the triarylamine donor (SGT-061, Figure 23).[177] With the increased π-system, the HOMO energy level was increased to -5.03 eV, and the device reached a PCE of 12.6%. A similar approach was taken by CHOU et al. in 2020 when they also used a linear motif and an alkyne bridge, but instead of the triphenylamine, they used a dialkyl-substituted aniline as donor group (YZT1 and YZT2, Figure 23).[178] The alkyl chains increased the solubility, and the length strongly impacted the molecular packing and hole mobility. The HOMO energy levels for both compounds were in the preferred range at -5.23 eV and -5.26 eV, respectively. YZT1 achieved a remarkable PCE of 13.1% without doping, and by doping with LiTFSI and TBP, the PCE could be increased to 14.4%. In 2021, SUN *et al.* published a broad overview of properties when exchanging metals in ZnP by five other first-row transition metals with theoretical and experimental results.[179] The HOMO energy level was similar, between -5.38 eV and -5.30 eV.

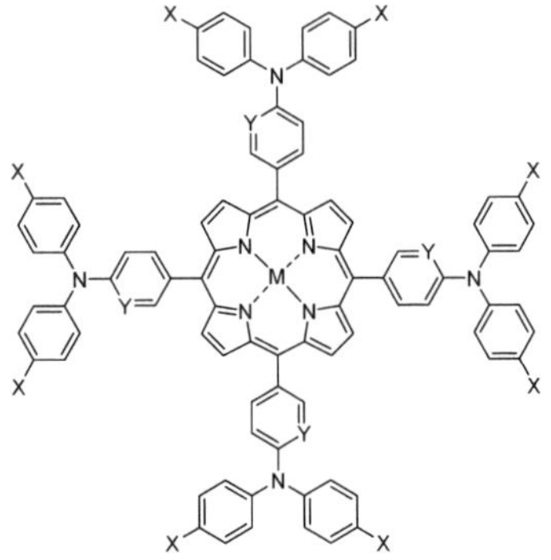

CuP: M = Cu²⁺, X = OMe, Y = C
ZnP = PZn-TPA-O: M = Zn²⁺, X = OMe, Y = C
PZn-TPA: M = Zn²⁺, X = H, Y = C
PZn-DPPA-O: M = Zn²⁺, X = OMe, Y = N
PZn-DPPA: M = Zn²⁺, X = H, Y = N

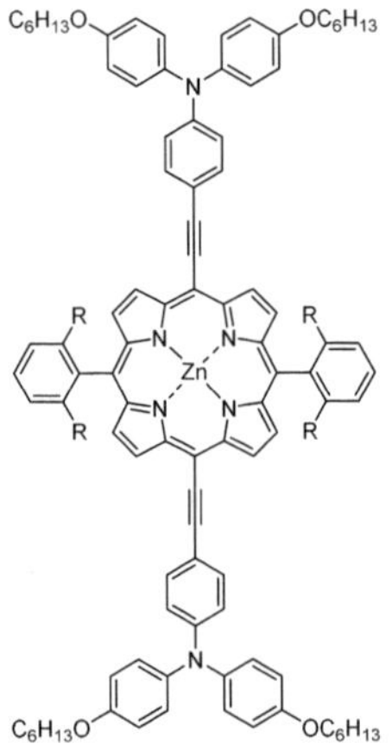

SGT-061, R = OC₈H₁₇

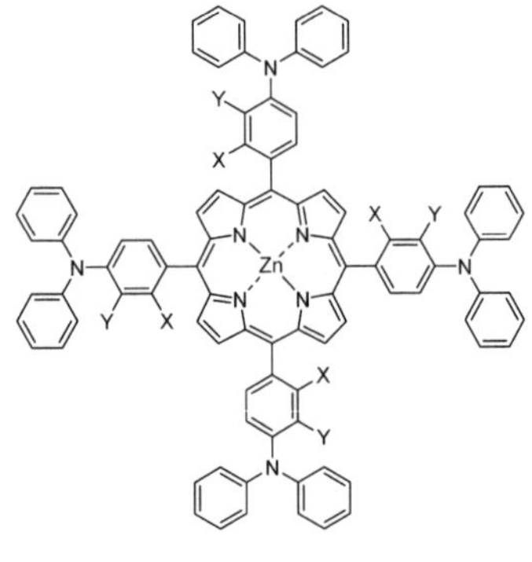

PZn-TPA: X = H, Y = H
PZn-2FTPA: X = F, Y = H
PZn-3FTPA: X = H, Y = F

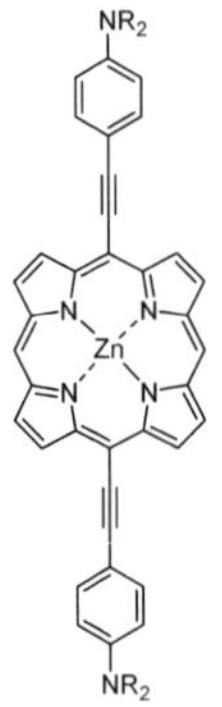

YZT1: R = butyl
YZT2: R = 2-butyl-octyl

Figure 23: Porphyrin-based organic HTMs.

2 AIM OF THE WORK

With the scarcity of fossil fuels and global temperature and greenhouse gas concentrations rising, the need for reliable and renewable energy sources is prevalent to allow the ongoing electrification of society while managing a turn-around/stop on global warming. The current gas crisis due to the war in Ukraine only highlights society's dependence on fossil energy sources, with prices for electricity and heating skyrocketing and massive new imports of natural gas and oil. As shown in the introduction, solar energy is one of the most promising, if not the most promising, renewable energy source, and research has already pro-vided incredible milestones. A good solar cell architecture has been developed with the perovskite solar cell. With KERASOLAR, an interdisciplinary collabora-tion of researchers at KARLSRUHE INSTITUTE OF TECHNOLOGY (KIT) has pledged to find more stable, efficient, lead-free perovskite alternatives for PSCs. Due to the wide variety of different possible electron transport materials and their sophisticated synthetical procedures, optimization of the overall PCE of perovskite solar cells is easier to achieve *via* the optimization of the hole transport materials as hole extraction, hole mobility and stability are levers that can greatly influence the performance of the devices. This work aims to synthe-size and thoroughly characterize possible organic hole transport materials to match the novel active layers. By molecular design, promising moieties were chosen, and by changing the end groups and π-bridges around different central building blocks, a library of HTMs is to be synthesized, and the effects of the substituents are to be analyzed. The most important characteristics of the newly synthesized materials are (i) suitable HOMO energy levels, (ii) a wide band gap to avoid electron extraction and recombination, (iii) the ability to build amor-phous thin-films, (iv) comparably high electron densities (v) high hole conduction and (vi) straightforward syntheses.

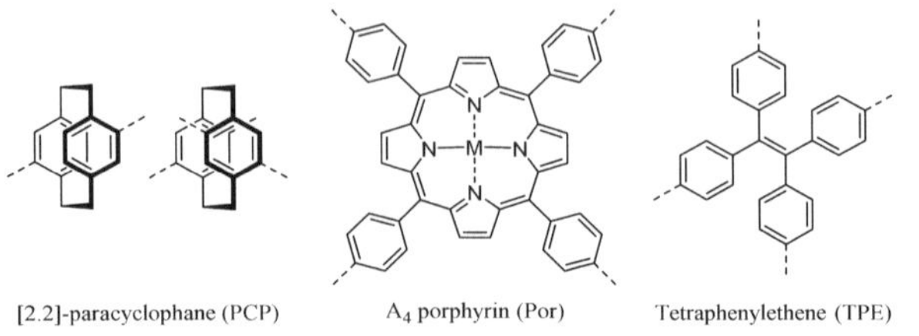

[2.2]-paracyclophane (PCP)   A₄ porphyrin (Por)   Tetraphenylethene (TPE)

Figure 24: Building blocks used as cores in this thesis.

For this work, the beforementioned building blocks [2.2]paracyclophane with a ps-*para* and bis(ps-*meta*)-para substitution pattern, a symmetrical A₄ porphyrin and tetraphenylethene (TPE) have been chosen for modification (Figure 24). A variety of donor groups, based on triphenylamines and carbazoles, will be synthesized, and together with a π-bridge (thiophene, EDOT, selenophene), these donor-π-systems will be attached to the different cores (Figure 25). The A₄ porphyrins offer another modification possibility as the coordinated metal can be removed and exchanged to evaluate its influence on the properties of the final hole transport material. As the [2.2]paracyclophane and the porphyrin already show interesting properties on their own with the through-space conjugation and the metal coordination, the straightforward molecular structure tetra-phenylethene has been chosen as the building block to determine the effect of methoxy substitution pattern on the triarylamine for the overall changes in HOMO and LUMO energy levels.

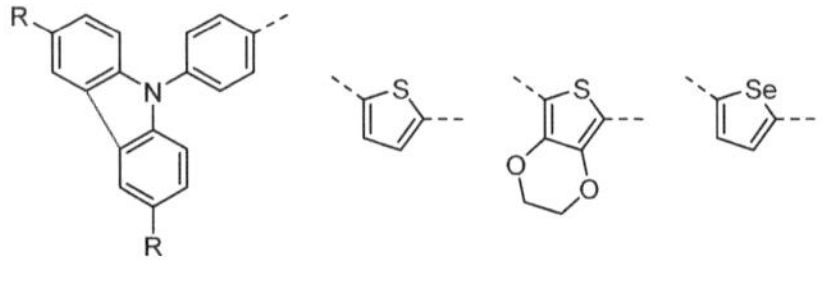

Figure 25: Building blocks used as donor- and π-moieties in this thesis.

3 MAIN SECTION

The main section will be divided into five different synthetical parts. The synthesis of the various donor molecules will be presented in the first part. The second part will show the different approaches to synthesizing the disubstituted [2.2]paracyclophanes and discuss their optoelectronic properties. The third block will provide insights into the extension of the π-system by adding two further substituents to the [2.2]paracyclophane core, and a comparison between the properties of the disubstituted and tetrasubstituted PCP will be made. A new core, the *meso*-tetrakis-phenyl-substituted porphyrin, is introduced in the fourth part. Apart from varying the different substituents, the porphyrin allows further modification by exchanging the coordinated metal inside the coordination site. Again, the optoelectronic properties of the different porphyrins will be compared. In the fifth block, the tetraphenylethene core was chosen as the central structure, and no π-bridges were used as connectors between the core and the donor groups. Using a simple core structure and no π-bridges, the effect of the methoxy-substitution pattern on triarylamine donor groups is investigated. Further results, such as theoretical insights, solubility, and solar cell performance, are provided and compared when available.

3.1 SYNTHESES OF THE DONOR MOIETIES

As shown in the introduction, triphenylamine and its partly locked counterpart, the carbazole, are excellent donor groups and have been successfully implemented into a wide array of hole transport materials. For this thesis, the donor groups were synthesized with different substituents and substitution patterns to determine their influence on the overall performance of the hole transport materials. The chosen substituents are *tert*-butyl groups (no mesomeric effect and a positive inductive effect), methoxy/methylthio groups (positive mesomeric effect and negative inductive effect), and the non-substituted donors as control groups. Substituents with negative mesomeric effect were not chosen as high electron density is elementary for good hole mobility. In addition to the electronic effect of the substituents, they should also increase or decrease the solubility of the compounds depending on the size and polarity.

3.1.1 Synthesis of the substituted Triarylamines

The literature-known, and well-investigated ULLMANN condensation was chosen to synthesize the triarylamines. It was first reported in 1903 by F. ULLMANN, who discovered improved yields for the coupling of phenols with aryl halides in the presence of copper and together with I. GOLDBERG, the method was applied to the reaction of aryl halides and amines under copper catalysis.[180-181] The copper catalyst's oxidation state is unknown, but radical processes are ruled out.[182] In 1999, GOODBRAND et al. could show that the addition of ligands improved the reaction time significantly, and a postulated mechanism is shown in Scheme 1.[183] During the first step, the copper(I) salt is coordinated by an added ligand, and then the aryl halide is bound to the catalyst by oxidative addition. In the second step, a ligand exchange between the halide and the amine takes place, and in the third step, the biarylamine is eliminated *via* the reduction of the copper catalyst. Typically, a base is added to the reaction mixture to neutralize the formed hydrohalide acids.

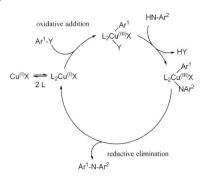

Scheme 1: A postulated mechanism for the ULLMANN condensation.

For the desired donor structures, two different pathways were possible. The substituted aryl halide could be coupled with aniline in the first step to afford the triarylamine, and in the second step, the phenyl ring belonging to the aniline could be brominated to activate the donor for the next step. This would allow the use of aryl bromines in the condensation reaction. But with the reactivity of the halides going from iodide with the highest reactivity to fluoride with the lowest reactivity, the use of substituted aryl iodide allowed for the use of 4-bromoaniline as a coupling partner and, therefore, rendered the bromination afterward redundant. In 2016, Li *et al.* published a procedure for the coupling of 4-iodoanisole with 4-bromoaniline under copper(I) chloride catalysis with 1,10-phenanthroline as ligand and potassium hydroxide as a base in refluxing toluene which yielded the desired 4-bromo-*N,N*-bis(4-methoxyphenyl)aniline (**1**) with excellent 91%.[184] The use of a DEAN-STARK apparatus additionally enforces product formation as it removes the water formed by the neutralization of hydrobromic acid with potassium hydroxide. This procedure was adapted, and with an increased reaction time from 16 h to 72 h, copper(I) chloride could be replaced with inexpensive and bench-stable copper(I) iodide without any impacts on the yield and the highest isolated yield for compound **1** being 93% (Scheme 2).

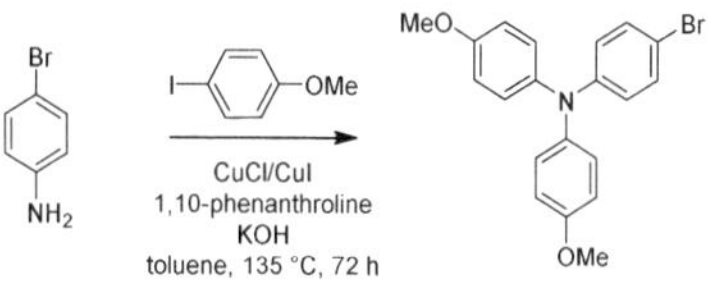

Scheme 2: Synthesis of triarylamine donors with compound 1 as an example.

Apart from the 4-iodoanisole, the 3-iodoanisole and the 2-iodoanisole were chosen as starting materials, and with the adapted procedure, the desired products 4-bromo-*N,N*-bis(3-methoxyphenyl)aniline (**2**) and 4-bromo-*N,N*-bis(2-methoxyphenyl)aniline (**3**) could be synthesized with excellent (95%) and fair yields (53%), respectively. The lower yield for compound **3** could be explained by the steric hindrance of the methoxy group during the reaction. The yield for compound **2** could be increased by 17% compared to similar procedures applied in literature, while for compound **3** the only published method applies copper(I) iodide nanoparticles and, therefore, is difficult to compare.[185-186] Next, two starting materials with more than one methoxy substituent were chosen, the 1-iodo-2,4-dimethoxybenzene and the 1-iodo-3,4,5-trimethoxybenzene which allowed the syntheses of 4-bromo-*N,N*-bis(2,4-dimethoxyphenyl)aniline (**4**) and 4-bromo-*N,N*-bis(3,4,5-trimethoxyphenyl)aniline (**5**). While the yield for compound **4** was slightly lower than the literature comparison (56% *versus* 67%), the yield for compound **5** was increased by more than three times compared to reporting literature (80% *versus* 26%).[187-188] In addition to the methoxy substituents, two further substituted triarylamines were synthesized, 4-bromo-*N,N*-bis(4-*tert*-butylphenyl)aniline (**6**) from 1-iodo-4-*tert*-butylbenzene and 4-bromo-*N,N*-bis(4-methylthiophenyl)aniline (**7**) from 4-iodo-thioanisole with very good yields (81% and 85% respectively). While the yield for compound **6** could be increased from 61% to 81%, the synthesis of compound **7** has not yet been described in literature *via* ULLMANN condensation.[189] An overview of the synthesized donor groups is provided in Figure 26.

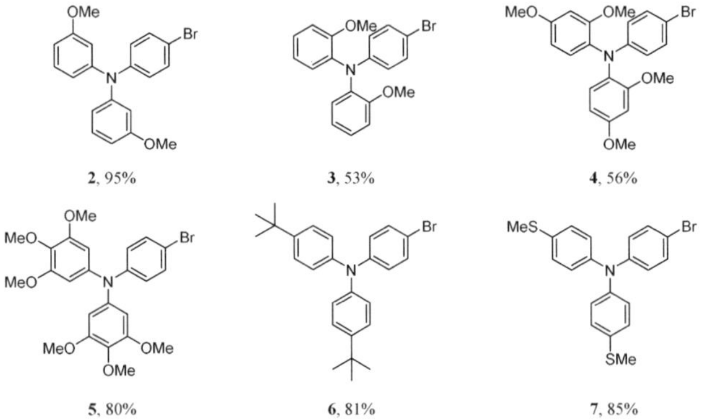

2, 95%   3, 53%   4, 56%

5, 80%   6, 81%   7, 85%

Figure 26: Overview of the synthesized triarylamine donors.

3.1.2 Synthesis of the substituted Carbazoles

Furthermore, two carbazole derivatives substituted with methoxy- and *tert*-butyl groups were synthesized. The carbazole as a general structure is interesting due to its higher rigidity than the triarylamine, as the linkage between the two phenyls prohibits twisting. The commercially available starting material 3,6-dibromo-9H-carbazole was chosen to synthesize the methoxy-substituted compound. With the addition of sodium methanolate and copper(I) iodide, the reaction followed the traditional ULLMANN-type Cu(I)-catalyzed cross-coupling of an alkoxide and the aryl bromide under high temperature and stochiometric amounts of sodium methanolate.[190] The procedure was modified from the publication of HOLZAPFEL *et al.* as the sodium methanolate was not synthesized *in situ,* but sodium methanolate in methanol was chosen as a reagent with the yield of 3,6-dimethoxy-9H-carbazole **8** matching the literature procedure (89% versus 92%).[191] In a second step, following the ULLMANN-type coupling, compound **8** reacted with 1,4-dibromobenzene *via* · copper(I) iodide catalysis yielding

38

9-(4-bromophenyl)-3,6-dimethoxycarbazole **9**. The reaction was performed according to a literature procedure, and the isolated yield was slightly lower at 78% compared to 90%.[192] A different approach was taken to synthesize 9-(4-bromophenyl)-3,6-ditert-butylcarbazole **10**. The commercially available 3,6-di*tert*-butyl-9H-carbazole was coupled following a literature procedure from WANG *et al.* with 1-fluoro-4-bromobenzene *via* activation of the C-F bond with cesium carbonate, which meant that no other metals were needed for this step and product **10** was isolated with an excellent yield of 90%.[193] An overview of the synthesized carbazoles is provided in Scheme 3.

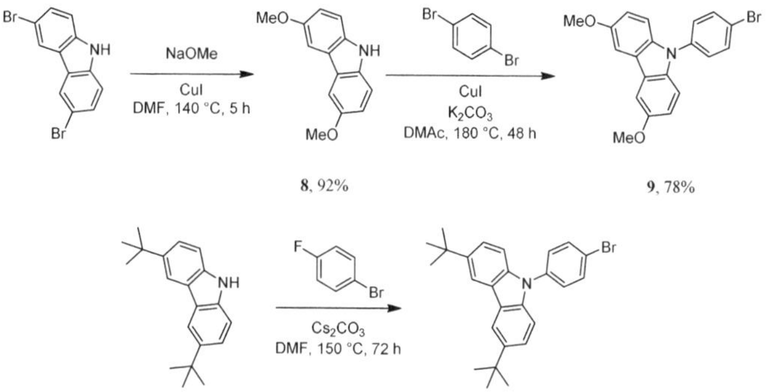

Scheme 3: Synthesized substituted carbazoles.

In conclusion, ten different donor groups were synthesized. For most reactions, the isolated yield was higher or similar to literature procedures, and by increasing the reaction time for the ULLMANN condensation, copper(I) iodide could be used as a catalyst compared to copper(I) chloride. The usage of the resulting donor moieties in reactions with the different targeted core building blocks will be discussed in the following chapters.

3.2   PSEUDO-*PARA*-SUBSTITUTED [2.2]PARACYCLOPHANES

Before discussing the synthesis of the desired substituted [2.2]paracyclophanes, a short explanation of the nomenclature is needed. As the [2.2]paracyclophane consists of two cofacially stacked benzene rings, the standardized nomenclature of *ortho*, *meta*, and *para* substitution pattern is extended by the substitution pattern on the second benzene ring, in which the prefix pseudo (ps) is added.

Therefore the following nomenclature is implemented: pseudo-*geminal*, pseudo-*ortho*, pseudo-*meta,* and pseudo-*para* (Figure 27a).[160] Additionally, Figure 27b shows the numbering of different carbon atoms in the structure.[194]

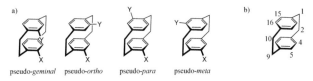

Figure 27: a) Nomenclature for disubstituted [2.2]paracyclophanes. b) Numbering of the [2.2]paracyclophane core.

3.2.1 Synthesis and Characterization

Two approaches were undertaken to synthesize the desired end products (Scheme 4). For the first approach, the desired molecule was synthesized from the core, the [2.2]paracyclophane, by first adding the π-bridge, and in a second step, the donor groups were coupled to yield the final product (inside-out approach). For the second approach, the donor moiety and the π-bridge are coupled in the first step, and in the second step, the donor-π molecules are added to the core (outside-in approach).

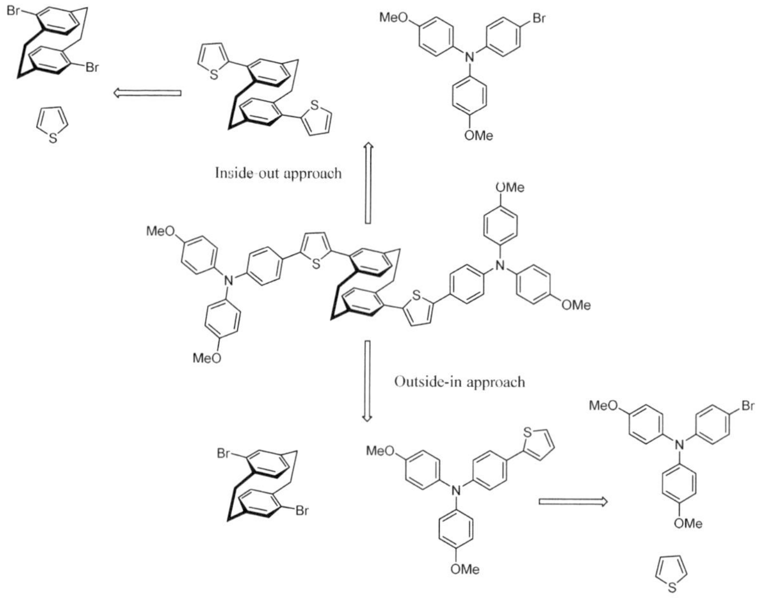

Scheme 4: Different synthetic approaches for synthesizing the desired target materials.

3.2.1.1   Inside-out approach

For the inside-out approach, several publications in literature have already described methods to synthesize the desired 4,16-dithienyl-[2.2]paracyclophane **11**. SALHI *et al.* described the synthesis of **11** *via* STILLE coupling where they used the thienyl stannane, the 4,16-dibromo-[2.2]paracyclophane and palladium as catalyst without reporting a yield but the mono-coupling of thienyl stannane with 4-bromo-[2.2]paracyclophane yielded 58% reported by DIX *et al.* in 1999.[195-198] While the STILLE coupling has proven its worth in organic chemistry over the last 40 years, more recently, the negative aspects regarding the use of stannanes, such as its toxicity and the aggravated workup to remove residues, have strengthened the search for viable options. In 2019, WEILAND *et al.* published a synthesis of **11** using commercially available GRIGNARD reagent 2-thienyl magnesium bromide as part of a modified KUMADA coupling employing a palladium catalyst and yielding **11** with 87%.[199-201] This procedure was repeated in this work but with a significantly lower yield of 19%, possibly due to the unstable nature of the GRIGNARD reagent.

Besides using metalates such as the stannanes or the GRIGNARD reagents, another carbon-carbon coupling method has gained increasing interest in literature, the palladium-catalyzed CH activation.[202] Due to the strong acidity of the hydrogen in the 2-position of thiophene and its derivatives, the palladium can insert into the C-H bond.[203] LIU *et al.* published an overview of different palladium catalysts, ligands, bases, and solvents for the CH activation of the similar structure 3,4-ethylenedioxythiophene (EDOT) with brominated aryls. They concluded that the highest yields were achieved with palladium acetate (Pd(OAc)$_2$) as the catalyst, P(*m*-Tol)$_3$ as a ligand, cesium carbonate as a base, and toluene as solvent at a temperature of 110 °C.[204] With 4,16-dibromo-[2.2.]paracyclophane kindly being provided by Dr. CHRISTOPH ZIPPEL (Working group BRÄSE), these conditions were then applied to the synthesis of compound **11** with thiophene as an activated coupling partner. The reactions were performed in closed vials under argon with different amounts of thiophene ranging from 2.50 to 20.0 equivalents. No product formation was observable for 2.50, 5.00, and 10.0 equivalents; but when increased to 20.0 equivalents thiophene, the desired product **11** could be isolated with a yield of 42%. Due to the boiling point of 87 °C for thiophene, an extensive excess was needed to dissolve enough thiophene inside the reaction mixture. The excess of thiophene, however, increased the difficulty during the workup as several different side products formed during the reaction. Therefore, the low yield can be attributed to low conversion or losses during the purification. To prohibit these side reactions, additional investigations were carried out. First, the ligand was changed from P(*m*-Tol)$_3$ to P(Ph)$_3$, but no product formation was observed. Secondly, KIEFFER *et al.* described a method using P(Cy)$_3$ instead of P(*m*-Tol)$_3$ as a ligand, potassium carbonate as a base, and pivalic acid as an additive while decreasing the temperature from 110 °C to 100 °C for the coupling of thiophene with aryl halides.[152, 205] While the use of P(Cy)$_3$ did not result in the formation of the desired product, the combination of pivalic acid and P(Ph)$_3$ showed promising results at 100 °C, yielding **11** with 71% and fewer side products, which decreased the effort needed during the workup (Scheme 6a). The use of pivalic acid as an additive has been described in the literature, and a proposed mechanism to explain the better yields is given in Scheme 5.[206] The aryl bromide is added *via* oxidative addition, and at the same time, the pivalic acid is deprotonated by potassium carbonate. The deprotonated pivalic acid can then bond to the palladium species. When an aryl with an acidic proton nears the palladium, the pivalic acid facilitates the insertion of the palladium into the C-H bond by abstracting the acidic hydrogen. Both aryls are then bonded to the palladium by H elimination, and the carbon-carbon bond is formed *via* reductive elimination.

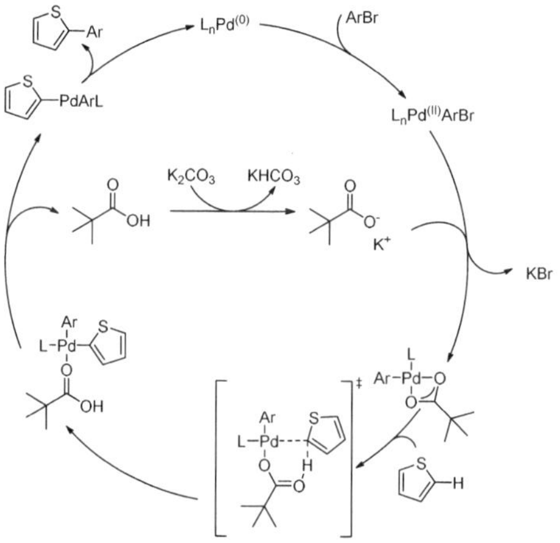

Scheme 5: Proposed mechanism with modification to fit the thiophene.[206]

Under these conditions, two more palladium catalysts, $Pd(PPh_3)_4$ and $Pd(dppf)Cl_2$, were tested, but both failed to produce the desired product. An overview of the different conditions is provided in Table 1.

Table 1: Overview of different CH activation conditions for the coupling of psp-Br₂-PCP with thiophene.

Entry	Thiophene (equiv)	Catalyst	Ligand	Additive	Temperature [°C]	Solvent	Yield [%]
1	2.50	Pd(OAc)₂	P(m-Tol)₃	-	110	Toluene	0
2	5.00	Pd(OAc)₂	P(m-Tol)₃	-	110	Toluene	0
3	10.0	Pd(OAc)₂	P(m-Tol)₃	-	110	Toluene	0
4	20.0	Pd(OAc)₂	P(m-Tol)₃	-	110	Toluene	42
5	20.0	Pd(OAc)₂	P(Ph)₃	-	110	Toluene	0
6	20.0	Pd(OAc)₂	P(Cy)₃	PivOH	110	DMF	0
7	20.0	Pd(OAc)₂	P(Ph)₃	PivOH	100	DMF	71
8	4.00	Pd(OAc)₂	P(Ph)₃	PivOH	100	DMF	12
10	4.00	Pd(dppf)Cl₂	-	PivOH	100	DMF	0
11	4.00	Pd(dppf)Cl₂	-	-	100	DMF	0
12	4.00	Pd(PPh₃)₄	-	PivOH	100	DMF	0
13	4.00	Pd(PPh₃)₄	-	-	100	DMF	0

Following the good results for the CH activation of the thiophene with 4,16-dibromo-[2.2]paracyclophane, the second coupling was performed using the most efficient conditions (Table 1, entry 7). Without the issue of a low boiling point (87 °C for thiophene), triarylamine 1 was added with 4.00 equivalents to prohibit polymerization of the 4,16-dithienyl-[2.2]paracyclophane. The reaction was controlled *via* thin-layer chromatography (TLC) and stopped after 13 days, with multiple new yellow and strongly fluorescent spots forming. The crude product was purified multiple times *via* column chromatography, and in the end, the desired product 4,16-di(4-(2-thienyl)-*N,N*-bis(4-methoxyphenyl)aniline)-[2.2]paracyclophane (Di-*p*-OMeTPAThio-PCP, **TM_31**) could be isolated with a yield of 38% (Scheme 6b).

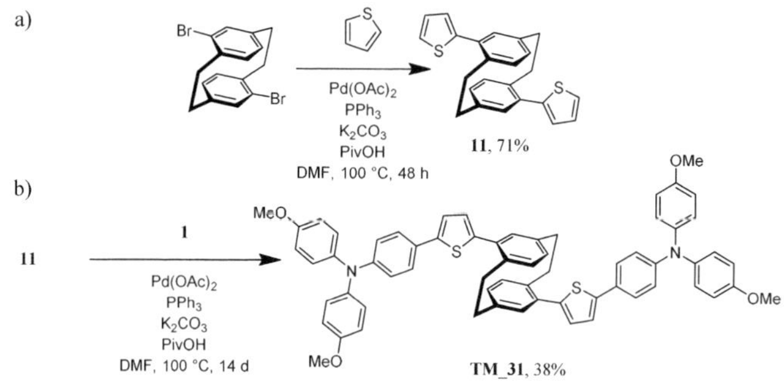

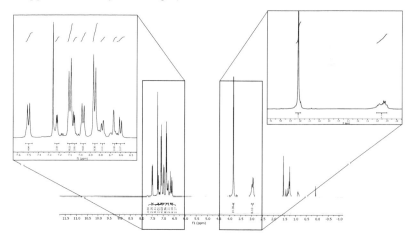

Scheme 6: a) CH activation of thiophene with 4,16-dibromo-[2.2]paracyclophane. b) CH activation of compound 1 and compound 11 to yield **TM_31**.

The purity of **Di-*p*-OMeTPAThio-PCP** (**TM_31**) was confirmed with ^{1}H NMR, ^{13}C NMR, and mass spectrometry. The ^{1}H NMR is depicted in Figure 28 and shows the typical signals expected for *para*-methoxy-substituted triaryl-amines (7.50 ppm, 7.10 ppm, 6.97 ppm, 6.86 ppm, and 3.82 ppm), the 2,5-sub-stituted thiophene (7.22 ppm and 7.06 ppm) and the pseudo-*para*-substituted [2.2]paracyclophane (6.78-6.60 ppm for the aromatic protons and 3.82 ppm and 2.94 ppm for the ethylene bridges).

Figure 28: ^{1}H NMR of compound **TM_31** in CDCl$_3$ via inside-out approach.

However, the synthesis could not be repeated successfully for any other donor groups, and with the strenuous workup and long reaction time, another strategy was essential to synthesizing a good library for further application.

3.2.1.2 Outside-in approach

The next approach was to build the desired products from the outside to prevent the formation of paracyclophane-containing side products, which were difficult to remove. This meant first coupling the donors with the π-bridges and, in a second step, performing the coupling with the [2.2]paracyclophane. The first synthetic approach used the same CH activation conditions described in the previous chapter. However, neither with nor without pivalic acid the desired donor-π moieties could be isolated, and the reaction control showed that most of the starting material **1** had defunctionalized. As a first step to increase the yield and prevent dehalogenation, the palladium catalyst was changed from Pd(OAc)$_2$ to the beforementioned Pd(PPh$_3$)$_4$, Pd(dppf)Cl$_2$, and in addition, PEPPSI™-iPr was investigated. Compared to the coupling of 4,16-dibromo-[2.2]paracyclophane with thiophene, the desired product formation was observed with all three catalysts when using 10.0 equivalents of thiophene. The isolated yields of 4-(2-thienyl)-N,N-bis(4-methoxyphenyl)aniline (**12**) ranged from 49% for Pd(dppf)Cl$_2$ over 56% for Pd(PPh$_3$)$_4$ to 75% for PEPPSI™-iPr after a reaction time of 45 hours (Scheme 7). The reaction could be reproduced with similar yields for the coupling of 4-bromo-N,N-bis(phenyl)aniline with thiophene to yield 4-(2-thienyl)-N,N-bis(phenyl)aniline (**13**) with up to 87% (Scheme 7) and therefor similar to literature.[207] A decrease in reaction time or equivalents of thiophene resulted in significantly lower yields (after 30 h reaction time yields lower than 40%).

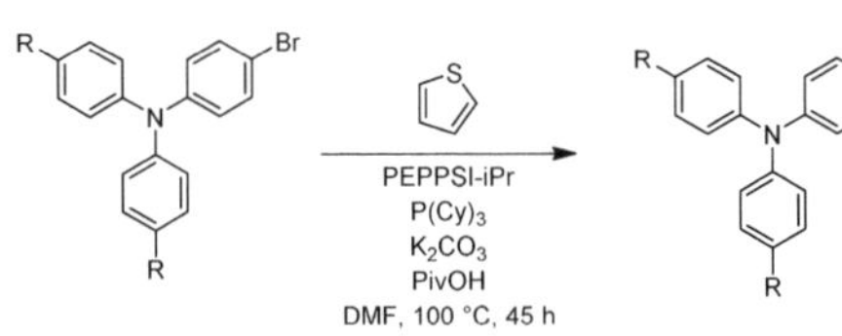

Scheme 7: CH activation of donors with thiophene.

Before applying this method to the other synthesized donor groups, it was tested on the other main π-bridge molecule of this work, the 3,4-ethylenedioxythiophene (EDOT). Due to the significantly higher boiling point of 193 °C, 1.50 equivalents of EDOT were sufficient to observe product formation. However, half of the starting material was dehalogenated before entering the reaction, and removing the unreacted EDOT from the product mixture was impossible with column chromatography or precipitation. Therefore slightly impure 4-(2-(3,4-ethylenedioxy)thienyl)-*N,N*-bis(4-methoxyphenyl)aniline **14** (established *via* NMR) and a 4-(2-(3,4-ethylenedioxy)thienyl)-*N,N*-bis(phenyl)aniline/EDOT mixture were isolated with 46% and <42% respectively (Scheme 8).

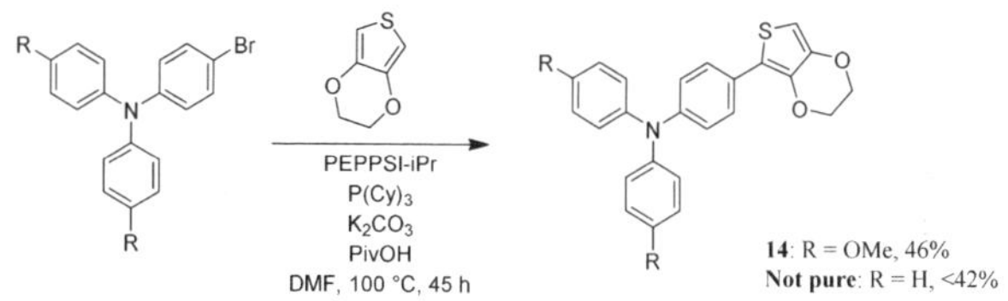

Scheme 8: CH activation of donors with EDOT.

Due to the significantly lower yields for the EDOT substitution pattern and the long reaction times for both CH activations, alternate synthetic routes were investigated. The most obvious method, using the commercially available thiophen-2-ylboronic acid with **1** in a SUZUKI-MIYAURA palladium-catalyzed cross-coupling reaction, did not show the expected result, as the boronic acid degraded during the reaction.[208] The more stable pinacol boronic ester was chosen as a reactant to prevent this decomposition. Following the typical conditions of a MIYAURA borylation, 2-pinacolborylthiophene **15** could be synthesized with a good yield of 76%.[209] A proposed mechanism of the MIYAURA borylation is depicted in Scheme 9a.[210] With a proposed mechanism of the SUZUKI-MIYAURA cross-coupling reaction shown in Scheme 9b, the similarities of both catalytic cycles are obvious.[211] The choice of base is crucial to prevent simultaneous borylation and cross-coupling. When using potassium acetate for the borylation, the reagents are not activated enough to undergo an additional cross-coupling, and the pinacolborane can be isolated.[210]

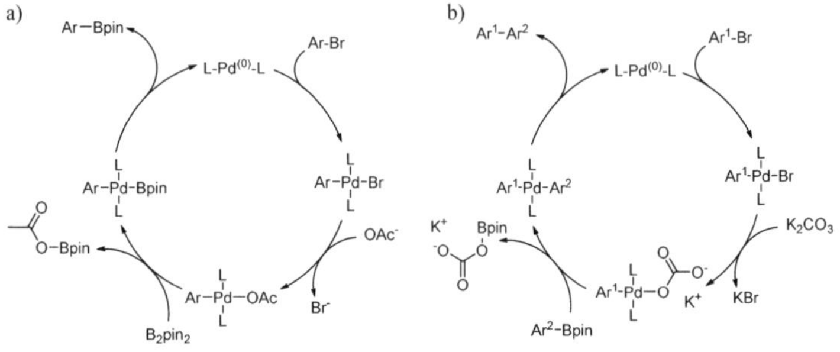

Scheme 9: Proposed mechanisms for a) the MIYAURA borylation and b) the SUZUKI-MIYAURA cross-coupling reaction.

However, the borylation of EDOT was the restricting factor as the increased reactivity for a second substitution led to the formation of mono- and disubstituted EDOT and a significant amount of unreacted starting material that could not be separated using conventional column chromatography or precipitation methods. Due to this synthetic hindrance, a second synthetic route was tested in which the acidic position of the thiophene is lithiated with *n*-butyllithium in the first step, and in a second step 2-isopropoxy-4,4′,5,5′-tetramethyldioxoborolane is added for the borylation following a procedure published by DIENES *et al.* in 2007.[212] 2-Pinacolborylthiophene **15** could be isolated with a very good yield of 87%, with a reaction time of only two hours. Using the same conditions, 2-pinacolboryl-3,4-ethylenedioxythiophene **16** and 2-pinacolborylselenophene **17** could be synthesized with 86% and 78% yield, respectively (Scheme 10). Selenophene was chosen due to its higher electron density and, therefore, might display better hole conduction.[213-214]

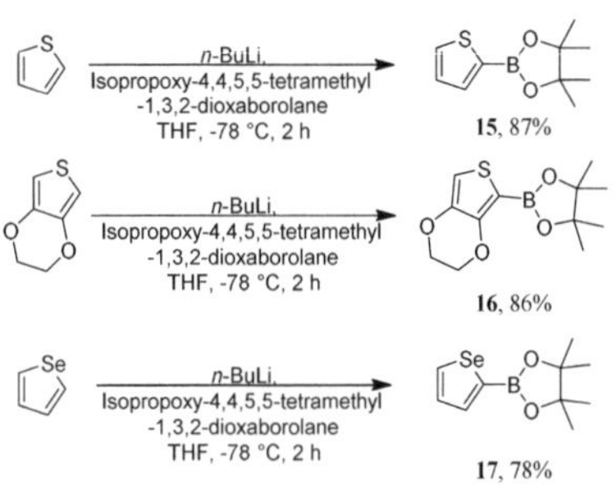

Scheme 10: Synthesis of borylated π-bridges.

With the borylated π-bridges successfully synthesized, the SUZUKI-MIYAURA palladium-catalyzed cross-coupling reactions with the donor groups were carried out under typical reaction conditions, particularly dimethyl sulfoxide as the solvent, Pd(dppf)Cl$_2$ as catalyst and potassium carbonate as the base (Scheme 11).

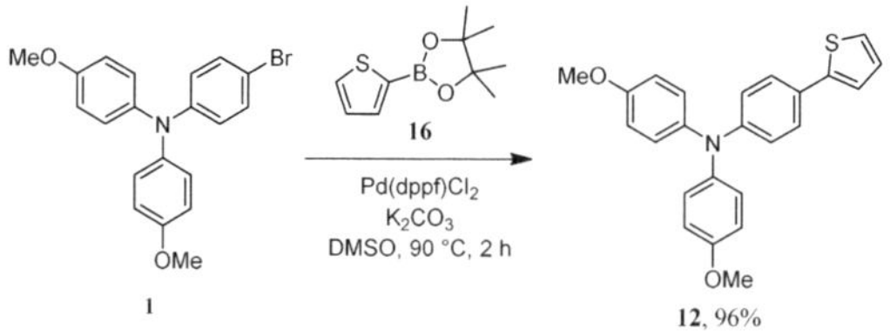

Scheme 11: SUZUKI-MIYAURA coupling of donor group and borylated π-bridge.

Only 1.20 equivalents of borylated π-bridge were necessary, and after a reaction time of two hours, yields between 71% and 98% were achieved for the triaryl-amine-based donors and thiophene and between 75% and 97% for the carbazole-based donors and thiophene. An overview of the synthesized donor-thiophene molecules is given in Figure 29. The synthesis of compound **12** has been described in various publications, and the yield of this method is comparable to other SUZUKI-MIYAURA couplings and 10% higher than similar STILLE couplings.[215] For compound **13**, the yield of 77% was lower than in STILLE couplings but the shortened reaction time of two hours compared to 24 hours makes up for the lower yield.[216] All other methoxy-substituted triarylamines have not yet been described in the literature (compounds **18**, **19**, **20**, and **21**), compound **22** has only been described *via* a different approach with lower yields, and **23** has only been described in a patent so far.[217] For the carbazoles as donor groups, compounds **24** and **26** have been described in literature before with similar yields to the presented yields in Figure 29.[218-219] Compound **25** has not yet been described in the literature, and summarizing, the reaction time for all synthesized compounds could be significantly decreased, in some cases, with similar yields.

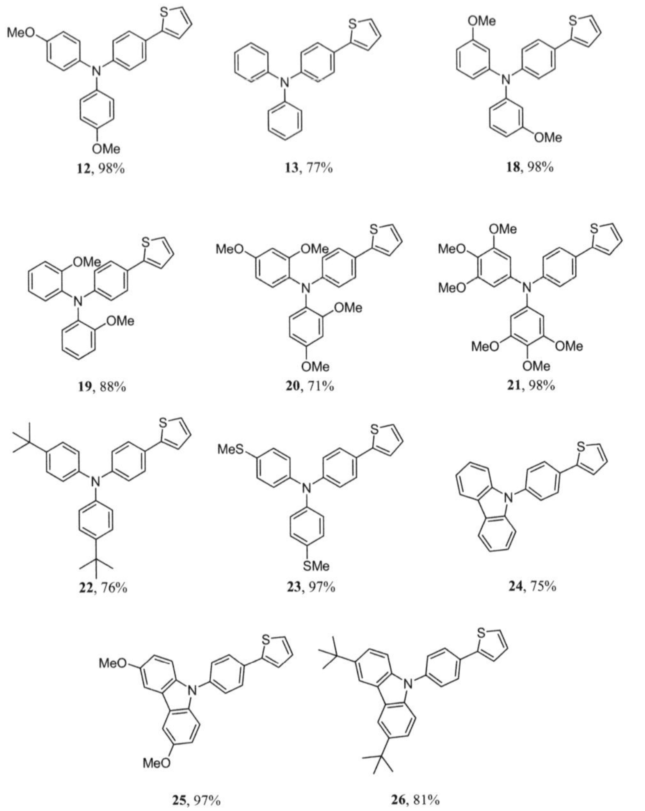

Figure 29: Overview of synthesized donor-thiophene molecules.

Apart from the cross-coupling with thiophene, similar molecules with EDOT and selenophene were synthesized (Figure 30). Compound **14** could be isolated with a similar yield to a published procedure following a STILLE coupling with significantly lower reaction time while exceeding published SUZUKI-MIYAURA conditions by 35%.[220] Here the significantly lower yield for compounds **27** and **28** are breaking the overall trend of yields over 70%. For compound **27**, the lower yield can be explained by reduced stability during the workup as the product seems to decompose on silica gel. For compound **28**, similar issues have surfaced due to the low polarity of the compound, the process of separating the side products by column chromatography was more challenging, and a certain

50

amount of inseparably mixed fractions were obtained that did not count towards the isolated yield. In contrast, the compound could be isolated with 79% in the literature.[152] Compounds **27**, **29,** and **30** have not yet been described in the literature.

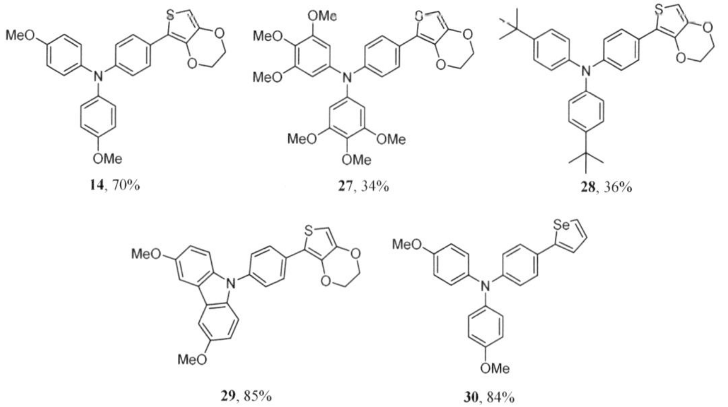

Figure 30: Overview of synthesized donor-EDOT and donor-selenophene molecules.

With a sufficient number of donor-π molecules synthesized, the last step for the outside-in approach was to connect the newly-made building blocks with the 4,16-dibromo-[2.2]paracyclophane core. As the π-bridges still possess an acidic proton on the 5-position, the first reaction type chosen to synthesize the target materials was another CH activation following the conditions already presented in the synthesis of compound **11**. Using palladium acetate as a catalyst with triphenylphosphine as a ligand and pivalic acid as an additive, Di-*p*-OMeTPAThio-PCP (**TM_31**) could be isolated with a yield of 62% after a reaction time of 18 hours (Scheme 12). Additionally, only 2.20 equivalents of the donor-π molecule **12** were needed to achieve this yield.

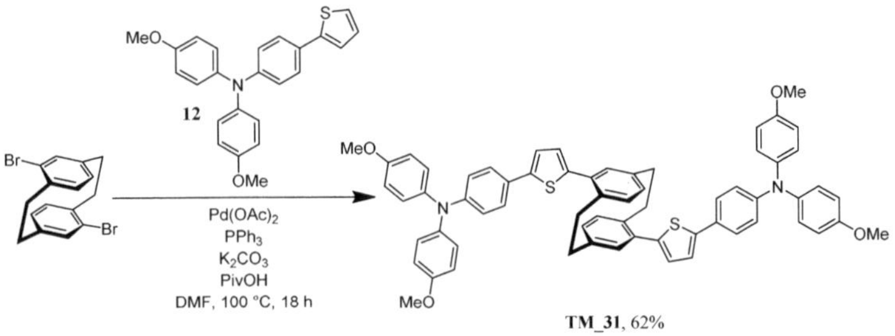

TM_31, 62%

*Scheme 12: Synthesis of target material **TM_31** via the outside-in approach.*

Apart from Di-*p*-OMeTPAThio-PCP (**TM_31**), this reaction could be repeated with various donor-π systems. An overview of the synthesized target materials is provided in Table 2.

Table 2: Overview of synthesized pseudo-para-substituted [2.2]paracyclophane target materials.

Entry	Starting material	Target material	Yield [%]
1	12	Di-*p*-OMeTPAThio-PCP (**TM_31**)	62
2	13	DiTPAThio-PCP (**TM_32**)	60
3	14	Di-*p*-OMeTPA-EDOT-PCP (**TM_33**)	37
4	21	Di-*mmp*-OMeTPAThio-PCP (**TM_34**)	48
5	22	Di-*p*-*tert*BuTPAThio-PCP (**TM_35**)	47
6	24	DiCbzThio-PCP (**TM_36**)	39
7	25	DiOMeCbzThio-PCP (**TM_37**)	32
8	26	Di-*tert*BuCbzThio-PCP (**TM_38**)	24

While the workup for all eight compounds was tedious, separating from other side products was easier than during the inside-out approach. As the PCP was only introduced in the second step and the thiophene only had one acidic proton left, the formation of chemically similar side products was reduced, and the main issue was fully removing the starting materials. The structures of all synthesized target materials are provided in Figure 31.

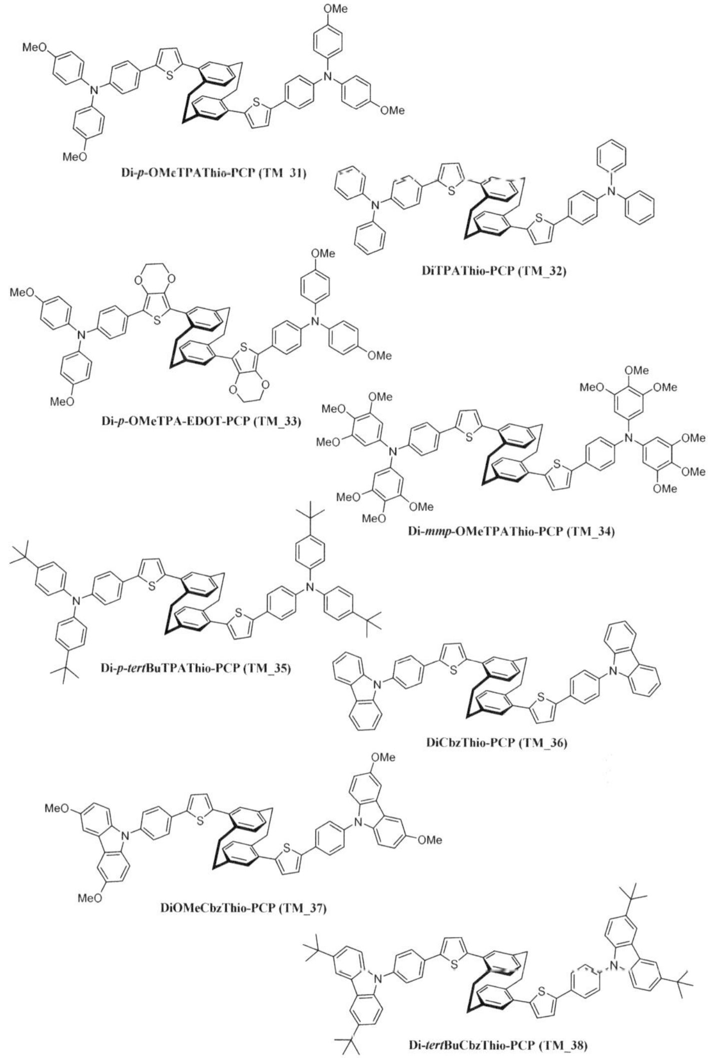

Figure 31: Overview of all synthesized disubstituted [2.2]paracyclophanes.

The products were purified *via* column chromatography on silica gel and dissolved in a mixture of cyclohexane and dichloromethane. The products precipitated by evaporating the dichloromethane under reduced pressure and could be isolated by centrifugation. All eight compounds are shades of yellow with a green fluorescence. When dissolved in slightly acidic solvents such as chloroform, the solution turned darker with time, and the ^{1}H NMR showed signs of degradation on all parts of the target materials.

The eight compounds were thoroughly characterized through ^{1}H and ^{13}C NMR and 2-dimensional heteronuclear single quantum coherence experiments (HSQC) so that all protons and carbons could be assigned. Additionally, high-resolution ESI MS spectra were measured, confirming the products' formation.

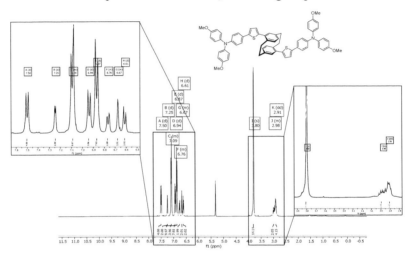

*Figure 32: ^{1}H NMR of **TM_31** in CD$_2$Cl$_2$ via outside-in approach.*

The depicted ^{1}H NMR of compound **TM_31** in Figure 32 was measured in deuterated dichloromethane, which explains the minor shifts compared to the ^{1}H NMR of **TM_31** *via* the inside-out approach (Figure 28). The ^{1}H NMR of **TM_31** *via* the outside-in approach shows the same characteristic peaks, but in comparison to the inside-out approach, the aromatic region shows much stronger and more distinct signals, confirming that the workup performed better in the outside-in approach. Additionally, the pattern of the peaks from the ethylene bridges confirms the substitution on the 4- and 16-position of the [2.2]para-cyclophane (see Figure 27b). The protons of positions 1 and 9 belong to the peak at 2.91 ppm, while the protons of positions 2 and 10 are split up. Due to the sp^3 hybridization of the carbons and no possible rotation, the chemical environment between the two protons H$_a$ and H$_b$ is different, and the proximity of the substituent R to proton H$_a$ leads to a stronger shift of the signal (in this case, 3.82 ppm for H$_a$ and 2.98 ppm for H$_b$, Figure 33). These typical and structure-defining peaks were observed in the eight synthesized target materials.

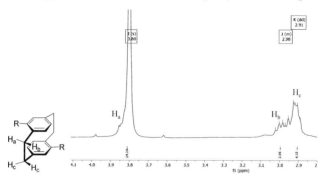

Figure 33: Difference between the protons on the ethylene bridge of 4,16-substituted-[2.2]paracyclophane.

In conclusion, eight different and literature-unknown target materials were synthesized, with variations to the end groups (H, OMe, *tert*-butyl), the donor groups (triphenylamine, carbazole), and the π-bridge (thiophene, EDOT). All materials were thoroughly characterized *via* NMR, mass, and IR, and their structure and purity were confirmed. By implementing the outside-in approach, the yields of the individual steps could be increased, and the workups were simplified. Nine synthesized donor-π groups have not yet been described in the literature, and the given literature yield was optimized for all other compounds while decreasing the reaction time. Further analysis of the optoelectronic properties of the synthesized target materials will be provided in the next chapter.

3.2.2 Optical and Electrochemical Characterization

After the structure of the eight target materials could be confirmed, other properties of the materials were investigated. To discuss the effects of the donor- and π-moieties, the target materials will be split into two groups depending on the donor groups (TPA and Cbz). The normalized absorbance (UV-vis) and emission (photoluminescence) spectra of all triarylamine-containing target materials are visualized in Figure 34.

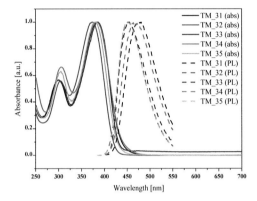

Figure 34: Normalized UV-vis spectra (in THF, lines) and normalized photoluminescence spectra (in THF, strokes) of the triarylamine-containing target materials.

The qualitative UV-vis measurements (Figure 34, abs) were performed in THF, and the maximum absorbance wavelength of each target material was used for the photoluminescence measurements (Figure 34, PL). The spectra show two main absorbance wavelengths for all five triarylamine-containing target materials. The absorbance at around 300 nm can be attributed to π→π* transitions, most probably localized in the donor moiety.[221] For this absorbance, no trend between the different substituents is observable as **TM_31** with *para*-methoxy-substitution and thiophene as π-bridge shows the strongest blue shift in absorbance at 297 nm similar to **TM_33** with *para*-methoxy substitution and EDOT as π-bridge (298 nm). This could additionally support the hypothesis of the transition localized in the donor moieties as there is no observable influence of the thiophene and EDOT. Expectedly, the absorbance of the tri-methoxy-substituted **TM_34** is redshifted by 9 nm due to more methoxy groups. The absorbances of the unsubstituted **TM_32** and the *tert*-butyl-substituted **TM_35** of 305 nm and 304 nm, respectively, do not follow a specific trend regarding substitution patterns. However, when investigating the maximum absorbance, the bands of the different target materials follow an observable trend. The absorbance around 370 nm and 400 nm can be attributed to intramolecular charge transfers (ICT) and the excitation from the HOMO to the LUMO.[165, 222]

The unsubstituted triarylamine **TM_32** shows the strongest blue shift with a λ_{max} at 373 nm. The second strongest blue shift can be observed for **TM_35** as the electron-donating capability of the *tert*-butyl group is higher than that of an unsubstituted triarylamine. The three absorbance maxima of the methoxy-containing compounds **TM_31** (Di-*p*-OMeTPAThio-PCP), **TM 33** (Di-*p*-OMeTPA-EDOT-PCP), and **TM_34** (Di-*mmp*-OMeTPAThio-PCP) are slightly redshifted compared to **TM_35** with λ_{max} going from 383 nm for **TM_34** over 384 nm for **TM_31** to 386 nm for **TM_33**. The similar λ_{max} for **TM_34** and **TM_31** can be attributed to the methoxy groups, and it seems like the number of methoxy groups does not influence the absorbance. However, the difference of 3 nm between **TM_31** and **TM_33** could be due to the change of the π-bridge. For the photoluminescence of the target materials, the two compounds with methoxy groups and thiophene as π-bridges **TM_31** and **TM_34** display the strongest red shift in emission with 482 nm and 467 nm, respectively. Interestingly, the *para*-methoxy-substituted **TM_33** with an EDOT-π-bridge displays a very similar maximum emission wavelength as the unsubstituted **TM_32** and the *tert*-butyl-substituted **TM_35** with 455 nm *versus* 452 nm and 450 nm respectively as no difference can be observed while the substitution patterns are different. Regarding the Stokes shift, the largest difference between absorbance maximum and emission maximum was recorded for **TM_31** with 98 nm, while the Stokes shifts of the other triarylamine-containing target materials were between 69 nm and 84 nm.

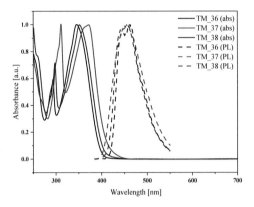

Figure 35: Normalized UV-vis spectra (in THF, lines) and normalized photoluminescence spectra (in THF, strokes) of the carbazole-containing target materials.

The results of the UV-vis and photoluminescence measurements of the carbazole-containing target materials are visualized in Figure 35. Similar to the triarylamine-containing target materials, all three absorbance spectra show two distinctive bands. Again, the absorbance at around 300 nm can be attributed to $\pi\rightarrow\pi^*$ transitions, while the absorbance around 360 nm corresponds to intramolecular charge transfers (ICT).[165, 221] The trend following the substitution pattern is similar for both absorbance bands. In both cases, the unsubstituted **TM_36** has the strongest blue shift (292 nm and 345 nm), while the *tert*-butyl-substituted **TM_38** is slightly red-shifted (297 nm and 351 nm), and the methoxy-substituted **TM_37** shows the strongest red shift with a first absorbance at 311 nm and a λ_{max} at 372 nm. Regarding photoluminescence, all three target materials display a similar maximum emission wavelength at 455 nm. This leads to a Stokes shift ranging from 83 nm for **TM_37** over 104 nm for **TM_38** to 110 nm for **TM_36**.

From these measurements, the optical band gap energy (ΔE_g) can be calculated (Equation 3).[152] The wavelength needed for calculating λ_{Eg} is measured at the intersection of the absorbance and photoluminescence spectra.

$$\Delta E_g = \frac{1240}{\lambda_{Eg}}$$

(3)

Equation 3: Calculation of the optical band gap.

An overview of all-optical results is provided in Table 3. The absorbance wavelengths of under 400 nm indicate that none of the materials would compete for photons with the perovskite active layer (absorbance between 400 nm and 700 nm), and due to the large Stokes shifts, the potential emission should not trigger any self-absorbance during the process. In addition, the large Stokes shifts imply that the molecules can undergo geometrical changes during excitation.[223] This could benefit possible defects in morphology in the devices. When comparing the different donor moieties, the maximum absorbance wavelengths of the carbazoles are strongly blue-shifted with a difference of around 30 nm for the unsubstituted **TM_32/TM_36** and the *tert*-butyl-substituted **TM_35/TM_38**. For both methoxy-substituted compounds, the absorbance maximum is more similar to λ_{max} of the triarylamine-containing **TM_31** at 384 nm and λ_{max} of the carbazole-containing **TM_37** at 372 nm. However, the effect of substituting the donor moieties remains similar for the carbazole and the triarylamine, as, in both structures, the methoxy-substituted target materials provide a stronger red shift than the *tert*-butyl-substituted target materials.

Table 3: Overview of optical properties for disubstituted [2.2]paracyclophanes.

Entry	Compound	λ_{max} [nm]	λ_{em} [nm]	Stokes shift [nm]	ΔE_g [eV]
1	TM_31	384	482	98	2.88
2	TM_32	373	452	79	2.96
3	TM_33	386	455	69	2.94
4	TM_34	383	467	84	2.90
5	TM_35	381	450	69	2.97
6	TM_36	345	455	110	3.01
7	TM_37	372	455	83	3.03
8	TM_38	351	455	104	3.04

The optical band gap of all synthesized target materials is similar, ranging from 2.88 eV to 3.04 eV. This means that the band gap of all target materials is large enough to prohibit electron conduction if the HOMO energy level of the materials matches the perovskite.

To experimentally determine the HOMO and LUMO energy levels of the synthesized compounds, cyclic voltammetry (CV) was conducted for all eight compounds. Again, for better comparison and overview, the triarylamine-substituted target materials will be discussed first. All cyclic voltammograms were measured in N_2-saturated DMF (0.1 M $[nBu_4N][PF_6]$) with a 3 mM target compound concentration and a scan rate of 100 mV s^{-1}. The cyclic voltammograms of the triarylamine-substituted target materials are visualized in Figure 36 for the oxidation process and Figure 37 for the reduction process.

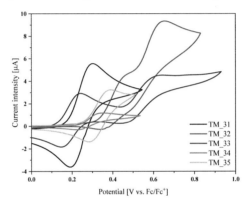

Figure 36: Oxidative cyclic voltammograms of the triarylamine-substituted target materials.

For all five compounds, oxidation processes are observable from 0.15 V to 0.8 V (Figure 36). For the target materials **TM_31**, **TM_33**, **TM_34**, and **TM_35**, the first oxidation processes are fully reversible, while the first oxidation process of **TM_32** shows a quasi-reversible behavior. In the range of the solvent, the compounds **TM_32** and **TM_33** display a second oxidation process, irreversible for **TM_32** and quasi-reversible for **TM_33**. The measured oxidation potentials *versus* the reference system ferrocene/ferrocenium are provided in Table 4.

The corresponding reductive cyclic voltammograms are visualized in Figure 37. For all compounds, the first reduction process was measured between -2.3 V and -2.8 V. All processes are quasi-reversible for the synthesized compounds, and an additional one or two reduction processes are detectable in a range of -0.5 V after the first reduction process. The measured reduction potentials *versus* the reference system ferrocene/ferrocenium are provided in Table 4.

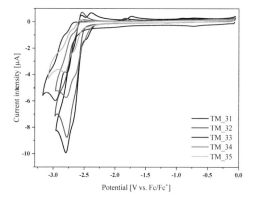

Figure 37: Reductive cyclic voltammograms of the triarylamine-substituted target materials.

The oxidative cyclic voltammograms of the carbazole-substituted target materials are visualized in Figure 38. All three compounds display oxidation processes between 0.4 V and 0.8 V. While the oxidation process for **TM_37** is quasi-reversible, the oxidation processes of **TM_36** and **TM_38** are not reversible. The measured oxidation potential of all three compounds is presented in an overview in Table 4.

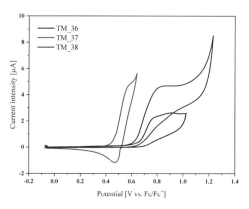

Figure 38: Oxidative cyclic voltammograms of the carbazole-substituted target materials.

The corresponding reductive cyclic voltammograms are visualized in Figure 39. Compared to the oxidative cyclic voltammograms of the carbazole-substituted target materials, the reduction processes all show a quasi-reversible character.

61

Depending on the material, the first observable reduction process is between -2.5 V and -2.8 V. Similar to the previously discussed reductive cyclic voltammograms of the triarylamine-substituted target materials, the carbazole-substituted target materials indicate second and third reductions shortly after the first reduction in a range of -0.2 V. The measured reduction potentials *versus* the ferrocene/ferrocenium reference system are displayed in Table 4.

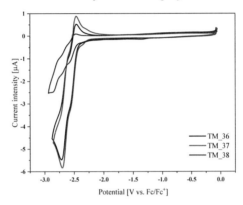

Figure 39: Reductive cyclic voltammograms of the carbazole-substituted target materials.

Apart from gaining insight into the redox properties of the compounds, the values gained from the oxidation and reduction potentials can be approximated to the HOMO and LUMO energy levels. For this, the following equation can be used:[224]

$$E_{HOMO/LUMO} = -(E_{ox}/E_{red} + 4.8)\ eV \qquad (4)$$

Equation 4: Calculating the HOMO and LUMO energy levels from the redox potential.

An overview of all obtained electrochemical values is provided in Table 4.

Table 4: Electrochemical properties of all synthesized di-substituted [2.2]paracyclophanes.

Entry	Compound	E_{ox} [V vs. Fc]	E_{red} [V vs. Fc]	E_{HOMO} [eV]	E_{LUMO} [eV]	$E_{HOMO-LUMO}$ [eV]	ΔE_g [eV]
1	TM_31	0.25	-2.59	-5.05	-2.21	2.84	2.88
2	TM_32	0.41	-2.58	-5.22	-2.23	2.99	2.96
3	TM_33	0.19	-2.69	-4.99	-2.11	2.88	2.94
4	TM_34	0.28	-2.51	-5.08	-2.30	2.78	2.90
5	TM_35	0.33	-2.60	-5.13	-2.20	2.93	2.97
6	TM_36	0.77	-2.51	-5.57	-2.30	3.27	3.01
7	TM_37	0.53	-2.51	-5.33	-2.29	3.04	3.03
8	TM_38	0.72	-2.53	-5.52	-2.27	3.25	3.04

The results in Table 4 provide some interesting and promising results. After calculating the approximate HOMO and LUMO energy levels from the CV measurements, the general idea behind the structural design is supported. All eight target materials contain exceptional energy levels while still being in the range for perovskite application. For the triarylamine-containing compounds, the overall deepest HOMO energy level (-5.22 eV) is achieved without substituting (**TM_32**, Table 4 entry 2). However, the deepest HOMO energy levels are achieved with all three carbazole-containing target materials (**TM_36**, **TM_37**, and **TM_38**, Table 4 entries 6, 7, and 8) with the deepest HOMO energy level at -5.57 eV for the unsubstituted carbazole-containing **TM_36**. Then there is an observable trend regarding the substitution with *tert*-butyl- and methoxy substituents. For both donor moieties (triarylamine and carbazole), the second deepest HOMO energy level is located in the *tert*-butyl-substituted target materials **TM_35** and **TM_38** with -5.13 eV and -5.52 eV, respectively (Table 4, entries 5 and 8). The shallowest HOMO energy level for the carbazole-containing target material is achieved by adding methoxy-substituents, and it is located at -5.33 eV. For the triarylamine-containing target materials, the three methoxy-substituted target materials show the shallowest HOMO energy levels. **TM_33** additionally possesses the EDOT building block as a π-bridge and has the shallowest HOMO energy level at -4.99 eV (Table 4, entry 3). In contrast, thiophene- and methoxy-containing target materials have remarkably similar

HOMO energy levels, with -5.05 eV for **TM_31** and -5.08 eV for **TM_34**. These values indicate that the substitution pattern on the donors and the π-bridges considerably impacts the energy level of the target materials. The shallower HOMO energy levels for the methoxy-substituted target materials indicate that stronger electron-donating substituents increase the HOMO energy levels similar to the exchange between thiophene and EDOT as π-bridges. The LUMO energy level is much less affected by the differences in the used building blocks. While the HOMO energy levels provide an overall range of 0.58 eV, the difference between the most and least stabilized LUMO energy level is only 0.19 eV. A comparison of the different band gaps furthermore confirms this. For the materials with deeper HOMO energy levels (**TM_32, TM_36, TM_37,** and **TM_38**, Table 4 entries 2, 6, 7, and 8), the band gaps are larger (>2.99 eV) than for the other materials (<2.93 eV). The comparison between the band gap calculated from the CV measurements and the optical measurements confirms the energetic structure of the eight target materials. The differences between the two measurement types are minor in most cases, with small deviations for **TM_34, TM_36,** and **TM_38**.

In addition to these experimental results, theoretical calculations regarding the properties were performed by DAVID ELSING (working group WENZEL, KIT). The calculated properties are presented in Table 5 and compared to the experimental values. Information on the methods can be found in the experimental section 5.1.

Table 5: Comparison between experimental and theoretical energy levels of disubstituted [2.2]paracyclophanes.

Entry	Compound	$E_{HOMO}^{calc.}$ [eV]	E_{HOMO}^{CV} [eV]	E_{LUMO}^{calc} [eV]	E_{LUMO}^{CV} [eV]
1	TM_31	-4.69	-5.05	-1.51	-2.21
2	TM_32	-4.99	-5.22	-1.68	-2.23
3	TM_33	-4.53	-4.99	-1.37	-2.11
4	TM_34	-4.79	-5.08	-1.65	-2.30
5	TM_35	-4.86	-5.13	-1.61	-2.20
6	TM_36	-5.42	-5.57	-1.92	-2.30
7	TM_37	-5.06	-5.33	-1.84	-2.29
8	TM_38	-5.28	-5.52	-1.88	-2.27

The values in Table 5 display a difference in values for both the HOMO and the LUMO energy levels. This deviation, however, was expected as the calculations were performed in the gas phase while the CV measurements were performed in solution. Due to the measurement in solution, intermolecular interactions can influence the energetic properties. Nonetheless, the trend regarding the HOMO energy levels is similar, the deviation for the HOMO energy levels is minimal, and the theoretical calculations confirm the experimental values. In conclusion, the influence of the different building blocks on the HOMO energy level can be summarized into three different trends that will be discussed with the different cores in the following chapters.

Carbazole < Triarylamine

H < *tert*-Butyl < Methoxy

Thiophene < EDOT

To summarize the results of this chapter, the most important values taken from the optical and electrochemical characterization are the absorbance wavelengths, the HOMO energy level, and the band gap. As shown in Table 3, the absorbance maxima are close to the ultraviolet light; therefore, the absorbance of all eight target materials should not negatively influence the performance of a solar cell. Next, the HOMO energy levels are between -5.0 eV and -5.6 eV and in the ideal range to support strong hole extraction and hole transport (VB of typical perovskites ~ -5.6 eV).[48, 225] Finally, the band gap of approximately 3.0 eV is large enough to prohibit electron extraction or conduction, as the CB of typical perovskites is located at around -4.0 eV.[48, 222] Therefore, the synthetic strategy worked well, and the synthesized materials were subjected to further testing regarding their applicability in PSCs. These results will be discussed in the next chapter.

3.2.3 Applicability of disubstituted [2.2]Paracyclophanes

With the promising results from the optical and electrochemical characterization, further investigation into the suitability was in order. As a first indicator, the thermal stability was measured *via* thermogravimetric analysis (TGA). Here, all eight compounds display stability to approximately 400 °C. For the compounds **TM_31** and **TM_35,** the first decrease of weight around 125 °C can be explained by the evaporation of some residue solvent (chapter 8.1, Figure 62).

After the thermal stability was verified, all eight compounds were delivered to ALEXANDER SCHULZ (working group COLSMANN, KIT) for further analysis. The following results were obtained by him and are presented after a thorough discussion to highlight the most interesting properties of the target materials.

First up, the question regarding general solar cell architecture had to be answered (see chapter 1.4.3). For this, **TM_31** was chosen as the test substance. Spin-coating from chloroform and chlorobenzene solutions worked well, but in the inverted architecture, the perovskite layer was subsequently added with DMF as the solvent, which dissolved the HTM layer. Therefore, the conventional setup was chosen. Next, the solubility of the target materials in chlorobenzene was investigated, and the materials were divided into three different categories. Firstly, the insoluble compounds with a solubility lower than 2 g L^{-1}, the sufficiently soluble compounds with a solubility between 2 g L^{-1} and 10 g L^{-1}, and the well-soluble compounds with solubilities higher than 10 g L^{-1}. The results are presented in Table 6.

Table 6: Results of the solubility tests for disubstituted [2.2]paracyclophanes.

Entry	Compound	Solubility in chlorobenzene [g L^{-1}]
1	TM_31	> 10
2	TM_32	2 < TM <10
3	TM_33	< 2
4	TM_34	> 10
5	TM_35	> 10
6	TM_36	< 2
7	TM_37	> 10
8	TM_38	> 10

From the obtained results, three compounds were disqualified from the further application (**TM_31**, **TM_33,** and **TM_36**, Table 6, entries 2, 3, and 6). For **TM_32** and **TM_36**, the lower solubility was not a big surprise as both compounds do not have any substitution on their donor moieties. The result was unexpected for **TM_33** (Di-*p*-OMeTPA-EDOT-PCP) as the methoxy-substitution should provide similar solubility to **TM_31**. This result could be explained by stronger stacking due to the EDOT groups, which provide additional anchoring points for hydrogen bonding.

For the next investigation, the ionization potential (IP) was measured from thin films *via* photoemission spectroscopy in air (PESA) and compared to the calculated values obtained by DAVID ELSING. The results are provided in Table 7.

Table 7: Comparison of the experimental and calculated ionization potentials.

Entry	Compound	Exp. IP [eV]	Calc. IP [eV]
1	TM_31	5.31	5.43
2	TM_32	5.63	5.77
3	TM_33	5.14	5.26
4	TM_34	5.31	5.57
5	TM_35	5.42	5.59
6	TM_36	5.86	6.20
7	TM_37	5.24	5.81
8	TM_38	5.73	6.00

The experimental and calculated ionization potentials provide a helpful expansion to the before discussed HOMO and LUMO energy levels. Again, the similarity between the experimental and calculated values is observable, and the deeper IP for the calculation is due to the gas-phase simulation. The experimental IP provides another clue regarding the suitability of the HTMs, as the thin-film measurements are closer to the energy levels inside the device than the measurements in the solution. The trend of ionization potentials is similar to the HOMO energy levels measured by cyclic voltammetry, with the unsubstituted donors containing the deepest ionization potentials (Table 7, entries 2 and 6) and the methoxy-substituted donors containing the shallowest ionization potentials (Table 7, entries 1, 3, 4 and 7).

With the newly obtained results, three target materials were chosen to be implemented as hole transport materials in perovskite solar cells. The target materials **TM_31** (Di-*p*-OMeTPAThio-PCP), **TM_35** (Di-*p-tert*-BuTPAThio-PCP), and **TM_37** (DiOMeCbzThio-PCP) convince with fitting HOMO and LUMO energy levels, high solubility and a large band gap. The structures of the three materials have been calculated and optimized *via* density functional theory (DFT) by DAVID ELSING, and the results are provided in Figure 40 (carbon in grey, oxygen in red, sulfur in yellow, nitrogen in blue). The comparison between **TM_31** (a) and **TM_37** (c) displays the difference between the TPA and the carbazole nicely, as the methoxy-substituted phenyls of the TPA unit can rotate independently while the carbazole blocks this rotation. The results of the device manufacturing are provided in chapter 3.6.1.

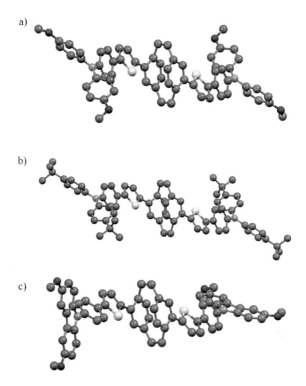

*Figure 40: DFT-optimized structures of a) **TM_31**, b) **TM_35**, and c) **TM_37**.*

3.3 BIS(PSEUDO-*META*)-*PARA*-SUBSTITUTED [2.2]PARACYCLOPHANES

Apart from choosing the almost linear pseudo-*para* substituted [2.2]para-cyclophanes to generally investigate the effects of the donors and π-bridges, the tetra-substituted bis(pseudo-*meta*)-*para*-[2.2]paracyclophane has been discussed as hole transport material in literature (see chapter 1.5.3.2).[163-165] The group of PARK *et al.* showed that a tetra-substitution is preferred compared to two triarylamines linked *via* one benzene unit and achieved PCEs of up to 18% for LiTFSI-doped Tetra-TPA-PCP. In 2021, LIN *et al.* published research in which they compared Tetra-TPA-(4,7,13,16)-PCP with the corresponding tetra-TPA-(4,5,15,16)-PCP, and they could provide insights that the latter pattern shows poorer characteristics as the 4,7,13,16-substituted-[2.2]paracyclophane provides better intermolecular charge transport due to its structure. They also displayed that implementing a bridge molecule increased the conjugation, and the resulting PCE of LiTFSI-doped WS-1 could be increased to 19% (Figure 21). These promising results and the knowledge from the previous chapter on disubstituted [2.2]paracyclophanes led to the investigation of different donor groups and π-linkers to better understand the structural properties and their effect on perovskite solar cell efficiency.

3.3.1 Synthesis and Characterization

While in literature, the tetrasubstituted [2.2]paracyclophanes were synthesized *via* SUZUKI-MIYAURA or HECK-MIZOROKI cross-coupling reactions, the first approach was to implement the beforementioned CH activation protocol for the tetra-(4,7,13,16)bromo-[2.2]paracyclophane, which was kindly provided by DR. CHRISTOPH ZIPPEL (working group BRÄSE).[226-227] Disappointingly, due to the increased steric hindrance, neither an increase in reaction time nor an increase in temperature and equivalents of the coupling partner could provide the successful synthesis of any of the target materials (Scheme 13a). In a second step, a borylation *via* lithiation of the donor-π moiety was investigated, and while the target mass was found with ESI MS, the target compound could not be isolated. The pinacolboronic ester probably defunctionalized during the workup (Scheme 13b). Therefore another strategy was in order. In 2014, KOBAYAKAWA *et al.* published a synthetic protocol using a NEGISHI cross-coupling reaction to connect two thiophene moieties with a pseudo-*ortho* iodinated [2.2]paracyclophane.[228-229] To obtain the required NEGISHI reactant, the thiophene was lithiated with *n*-butyllithium, then a zinc chloride solution was added, and with the NEGISHI reactant ready, the [2.2]paracyclophane was introduced to the reaction, and they could isolate the desired product with a yield of 85%. One postulated mechanism for the NEGISHI cross-coupling reaction is depicted in Scheme 13c.[230-231] Starting from a palladium catalyst with an oxidation state of 0, the halogenated starting material is bound *via* oxidative addition (1), thus oxidizing the catalyst to +II. In the second step, the zinc-containing coupling partner is added *via* transmetalation (2), and the corresponding zinc salt is removed from the cycle. In the third step, there can be a *cis/trans* isomerization to get the two aromatic systems into a closer vicinity (3), and in the last step, the desired product is formed *via* reductive elimination and thus reducing the catalyst back to an oxidation state of 0 (4).

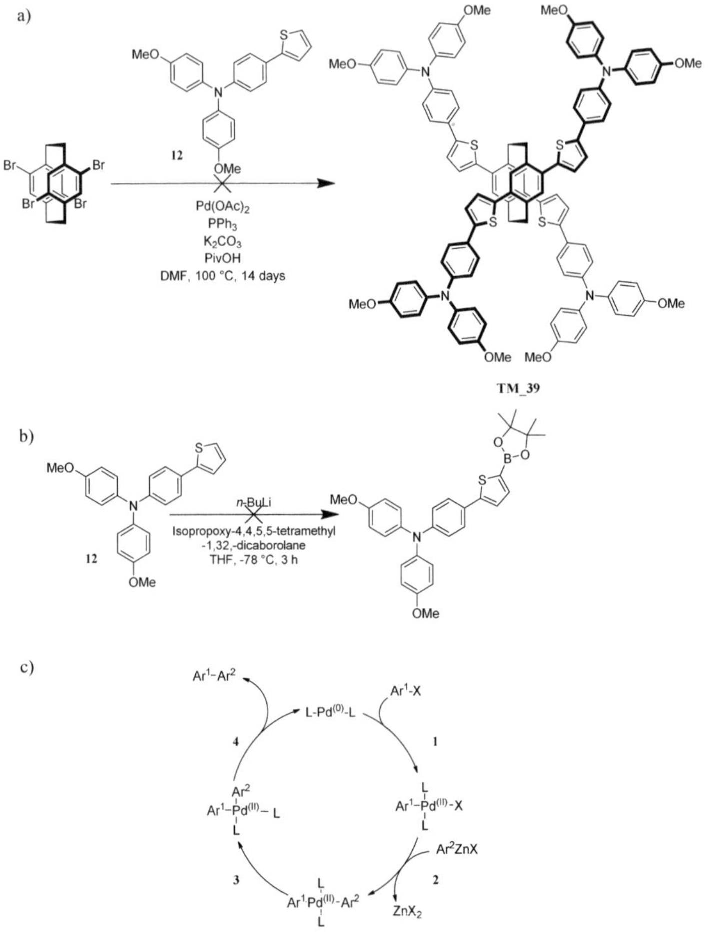

Scheme 13: a) Unsuccessful target material synthesis via CH activation. b) Unsuccessful borylation of compound 11. c) Postulated mechanism of the NEGISHI cross-coupling reaction.

The beforementioned synthetic protocol was modified slightly and then applied to synthesize compound **12** with the tetrasubstituted [2.2]paracyclophane. Compound **12** was lithiated, zinc chloride dissolved in THF (1.00 M) was added, and the NEGISHI reactant was added to a solution of tetra(4,7,13,16)-bromo-[2.2]paracyclophane and PEPPSI™-iPr. After stirring at 70 °C for 18 hours, the desired product could be purified *via* column chromatography and isolated with a yield of 64%.

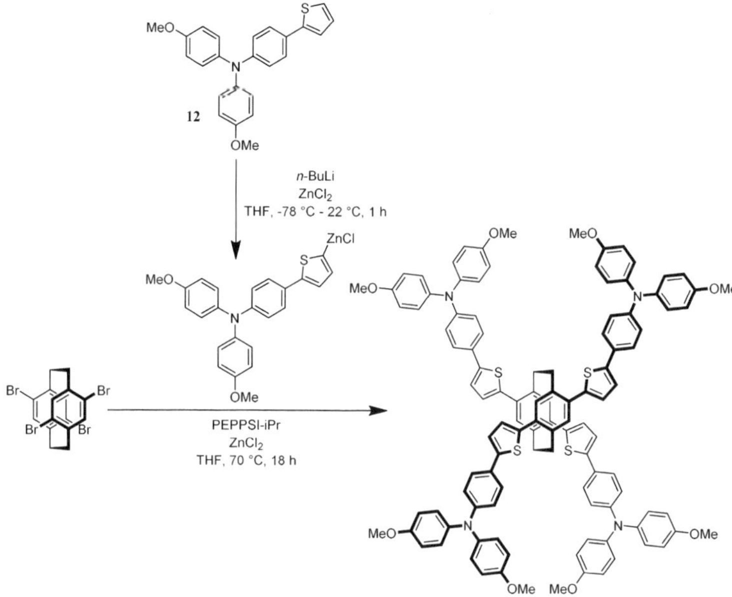

Scheme 14: Successful synthesis of **TM_39**.

As the coupling took place on all four brominated positions, this corresponds to a successful coupling rate of 89% on each of the sites. The successful synthesis of **TM_39** was confirmed with ^{1}H NMR, ^{13}C NMR, HSQC, MS, and IR. The ^{1}H NMR of **TM_39** is visualized in Figure 41. The overall symmetry of the compound is nicely depicted in the NMR spectrum, and all relevant signals are observable. The 16 protons of the four phenyl rings connecting the amine with the thiophene are divided into two signals of eight protons with shifts at 7.45 ppm and 6.88 ppm, while the protons of the eight methoxy-substituted phenyl rings are divided into two multiplets of 16 protons each. These signals are shifted to 7.06 ppm and 6.84 ppm. The protons of the four thiophene moieties are divided into two doublets of four protons each and shifted to 7.24 ppm and 7.14 ppm, while the remaining four protons of the two aromatic

73

rings of the [2.2]paracyclophane are shifted as a singlet to 7.00 ppm. This singlet is proof of the anticipated fourfold substitution in positions 4, 7, 13, and 16. The 24 protons of the methoxy groups are shifted to 3.78 ppm, and similar to the ^{1}H NMR of **TM_31**, some of the protons of the ethylene bridges are shifted similarly.

Compared to the di-substitution, the protons of the ethylene bridges do not demonstrate a similar splitting of the signals closer to the substituted positions, again providing proof that the desired molecule was formed. The second set of protons from the ethylene bridges is expectedly to shift to 2.98 ppm. In ESI MS analysis, the m/z of $[M]^+$ as well as $[M]^{2+}$ and $[M]^{3+}$ were found, and the high-resolution confirmed the exact mass of the product (m/z $[M]^+$ calcd for $C_{112}H_{92}N_4O_8S_4$, 1748.5798 and found 1748.5778).

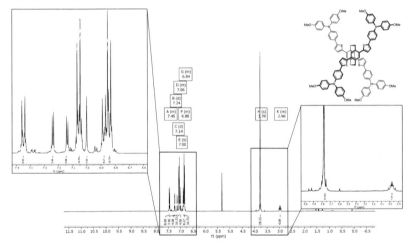

*Figure 41: ^{1}H NMR of **TM_39**, measured in CD$_2$Cl$_2$.*

Four additional tetra-substituted [2.2]paracyclophanes were successfully synthesized. With the insights taken from the thorough characterization of the di-substituted target materials, the compounds **13** (TPAThio), **14** (*p*-OMeTPA-EDOT), **22** (*p-tert*BuTPAThio), and **25** (OMeCbzThio) were chosen as starting materials as their optoelectronic properties were in the desired range. Compounds **13** and **14** were chosen despite the solubility of **TM_32** and **TM_33** as the two additional substituents could improve the solubility drastically. An overview of the synthesized target materials is given in Table 8.

Table 8: Overview of the synthesized tetra-substituted [2.2]paracyclophanes.

Entry	Starting material	Target Material	Yield [%]
1	12	Tetra-*p*-OMeTPAThio-PCP (**TM_39**)	64
2	13	TetraTPAThio-PCP (**TM_40**)	53
3	14	Tetra-*p*-OMeTPA-EDOT-PCP (**TM_41**)	12
4	22	Tetra-*p*-*tert*BuTPAThio-PCP (**TM_42**)	75
5	24	TetraOMeCbzThio-PCP (**TM_43**)	56

For the four target materials containing thiophene as π-bridge, the yield was between 53% and 75% (Table 8, entries 1, 2, 4, and 5), and therefore a conversion of 85% to 95% for each of the substituents. Similar to the di-substituted [2.2]paracyclophanes, the tetra-substituted [2.2]paracyclophanes were obtained as bright yellow solids with a greenish fluorescence. Regarding **TM_41**, the yield of only 12% stands out. Here the loss of target material could be attributed to two different causes. First, the EDOT-moiety as π-bridge needs more space than the thiophene due to its additional ethylene bridge, which may be the reason for a lower conversion as the third and fourth substitutions might be sterically blocked. Secondly, the lithiation of the donor-EDOT moiety might not work as well as for the thiophenes resulting in a lower concentration of active reagent and decreasing the yield. All five target materials were thoroughly characterized with ^{1}H NMR, ^{13}C NMR, HSQC, MS, and IR and while the target materials **TM_40**, **TM_42,** and **TM_43** displayed similar NMR spectra as **TM_39**, the ^{1}H NMR of **TM_41** showed some significant differences. The ^{1}H NMR of **TM_41** is visualized in Figure 42.

75

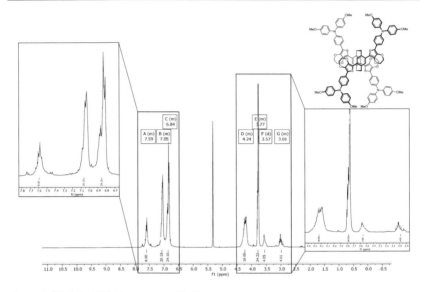

*Figure 42: ^1H NMR of **TM_41**, measured in CD_2Cl_2.*

While the signals are shifted into the expected areas, the resolution is notably worse compared to the ^1H NMR of **TM_39** (Figure 41). However, the structure-defining signals from the ethylene bridge of the EDOT moieties, the methoxy groups, and the ethylene bridges of the [2.2]paracyclophane are observable with the expected shifts. The signal of the 16 protons of the ethylene groups from the EDOT moiety is shifted to 4.24 ppm, while the signal of the methoxy protons is shifted to 3.77 ppm, and the eight PCP-bridge proton signals are shifted to 3.57 ppm and 3.01 ppm respectively. Due to the low boiling point of dichloro-methane and the insolubility of **TM_41** in other available deuterated solvents, no temperature-dependent NMR spectra could be measured to further confirm the rotational hindrance introduced by the EDOT moieties. However, the analysis of the ESI MS high-resolution spectrum confirmed the formation of the desired target material (m/z $[M]^+$ calcd for $C_{120}H_{100}N_4O_{16}S_4$, 1980.6017 and found 1980.6005).

To further represent the influence of the thiophene-π-bridge *versus* the EDOT-π-bridge, the DFT-optimized structures of **TM_39** and **TM_41** are presented in Figure 43 (carbon in grey, oxygen in red, sulfur in yellow, nitrogen in blue).

While for **TM_39,** the thiophene units can rotate alongside the bonds connecting them to the PCP and the TPA, the additional ethylenedioxy-subunits are hindering the rotation alongside the bonds and increasing the rigidity inside the molecule, resulting in the lower resolution of the NMR spectra.

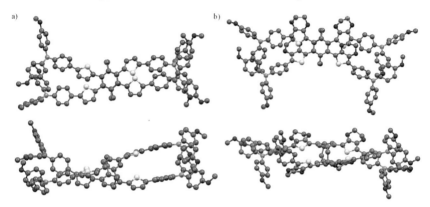

*Figure 43: DFT-optimized structures of a) **TM_39** and b) **TM_41**.*

In conclusion, all five desired, and literature-unknown target materials containing a tetra-substituted [2.2]paracyclophane core could be synthesized *via* a NEGISHI cross-coupling reaction with satisfying yields. The variation of π-bridges (thiophene, EDOT), donor moieties (triarylamine, carbazole), and end groups (H, OMe, *tert*-butyl) could provide more insights regarding their properties and will act as promising new materials in the hole transport materials library. All materials were thoroughly characterized *via* NMR, mass, and IR, and their structure and purity were confirmed. Further analysis of the opto-electronic properties of the synthesized target materials will be provided in the next chapter.

3.3.2 Optical and Electrochemical Characterization

With all necessary characterization done and the structure of the target materials confirmed, the characterization of the optical and electrochemical properties was completed. The UV-vis and photoluminescence spectra of the five target materials are visualized in Figure 44.

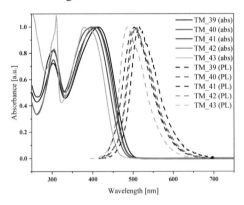

Figure 44: Normalized UV-vis spectra (in THF, lines) and normalized photoluminescence spectra (in THF, strokes) of the tetra-substituted [2.2]paracyclophane target materials.

The qualitative UV-vis measurements (Figure 44 abs) were performed in THF, and the maximum absorbance wavelength was used for the photoluminescence measurements (Figure 44 PL). Similar to the UV-vis measurements of the di-substituted [2.2]paracyclophanes (Figure 34 and Figure 35), the UV-vis spectra of the tetra-substituted [2.2]paracyclophanes display two main absorbance wavelengths. The first absorbance at around 310 nm is strongest for **TM_43**, the target material with the OMeCbzThio side groups. Similar to the di-substituted **TM_37**, the absorbance is stronger than the absorbance at higher wavelengths. The first absorbance wavelength is blue-shifted for the other four target materials containing triarylamine donors to between 301 nm for **TM_41** and 305 nm for **TM_40**. As with the di-substituted [2.2]paracyclophanes, this absorbance could be contributed to a $\pi \rightarrow \pi^*$ transition in the donor moiety.[221] The same trend is observable for the second absorbance as for the di-substituted [2.2]paracyclophanes. Regarding the triarylamine-containing target materials and thiophene as π-bridge, the unsubstituted **TM_40** absorbs best at 398 nm while the *tert*-butyl-substituted **TM_42** is red-shifted to 406 nm, and the meth-

78

oxy-substituted **TM_39** is red-shifted to 417 nm. Again, this additional redshift can be explained by the electron-donating strength of the different substituents, and the absorbance can again be attributed to an intramolecular charge transfer (ICT) between the donor groups and the core.[165] Compared to the triarylamine-containing target materials, the carbazole-containing **TM_43** demonstrates the strongest blue shift, with a maximum absorbance at 379 nm, again following the trend already seen at the di-substituted [2.2]paracyclophanes. However, when comparing the range of the absorbance maxima between the target materials with two substituents (see Table 3) and the target materials with four substituents (see Table 9), a red shift for all tetra-substituted target materials is observable. The smallest red shift is contributed for **TM_43** with 7 nm while the four tri-arylamine-containing target materials display larger red shifts between 25 nm for **TM_40** and **TM_42** and 33 nm for **TM_39**. Similar to the absorbance wavelengths, the maximum emission wavelengths of all five target materials contain a red shift compared to the corresponding di-substituted target materials. While the maximum emission wavelengths of the di-substituted target materials ranged from 450 nm for **TM_35** to 482 nm for **TM_31** (see Table 3) the maximum emission wavelengths of the tetra-substituted target materials ranged from 486 nm for **TM_43** to 517 nm for **TM_39**. Nonetheless, the trend is again similar, with **TM_31** and **TM_43** displaying the most red-shifted emission wavelengths. The Stokes shifts for all five target materials have increased by around 20 nm compared to the corresponding disubstituted target materials except for **TM_39**, in which the difference is only 2 nm. An overview of all optical results is provided in Table 9.

Table 9: Overview of optical properties for tetra-substituted [2.2]paracyclophanes.

Entry	Compound	λ_{max} [nm]	λ_{em} [nm]	Stokes shift [nm]	ΔE_g [eV]
1	TM_39	417	517	100	2.60
2	TM_40	398	505	107	2.67
3	TM_41	410	502	92	2.69
4	TM_42	406	508	102	2.65
5	TM_43	379	486	107	2.76

The optical band gap was calculated according to Equation 3. The red shifts for all five new tetra-substituted target materials could be attributed to the increased π-system and, therefore, smaller band gap between the HOMO and LUMO energy levels. This is supported by the obtained values from the optical band gaps. Compared to the optical band gaps of the di-substituted target materials of close to 3.00 eV, the newly calculated optical band gaps for the tetra-substituted target materials have decreased to values between 2.76 eV for **TM_43** and 2.60 eV for **TM_39**. Again, these band gaps are large enough to prohibit electron extraction and conduction from the perovskite.

To determine the synthesized compounds' experimental HOMO and LUMO energy levels, cyclic voltammetry was performed for all five compounds. The conditions for the cyclic voltammetry measurements are found in chapter 3.2.2. The oxidative cyclic voltammograms of the tetra-substituted [2.2]paracyclophanes are visualized in Figure 45, and the reductive cyclic voltammograms are visualized in Figure 46.

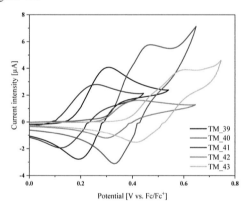

Figure 45: Oxidative cyclic voltammograms of the tetra-substituted [2.2]paracyclophanes.

The five synthesized target materials exhibit oxidation processes between 0.1 V and 0.6 V. While the oxidation processes for **TM_39, TM_41,** and **TM_42** are reversible, the oxidation processes of **TM_40** and **TM_43** display quasi-reversible behavior. The measured oxidation potentials *versus* the reference system ferrocene/ferrocenium are provided in Table 10.

All compounds exhibit multiple reduction processes in the range of the solvent. For all five compounds, the first reduction process is in the range of -2.26 V (**TM_43**) and -2.54 V (**TM_41**). Furthermore, the reduction processes for all five compounds are quasi-reversible. The measured reduction potentials *versus* the reference system ferrocene/ferrocenium are displayed in Table 10.

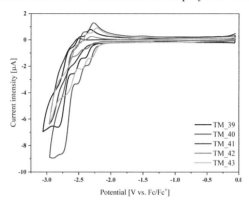

Figure 46: Reductive cyclic voltammograms of tetra-substituted [2.2]paracyclophanes.

In addition to the experimental values for the oxidation and reduction potentials, the energy levels of HOMO and LUMO were calculated according to Equation 4, and all obtained electrochemical values are presented in Table 10.

Table 10: Electrochemical properties of all synthesized tetra-substituted [2.2]paracyclophanes.

Entry	Compound	E_{ox} [V vs. Fc]	E_{red} [V vs. Fc]	E_{HOMO} [eV]	E_{LUMO} [eV]	$E_{HOMO-LUMO}$ [eV]	ΔE_g [eV]
1	**TM_39**	0.25	-2.37	-5.05	-2.43	2.62	2.60
2	**TM_40**	0.39	-2.32	-5.19	-2.48	2.71	2.67
3	**TM_41**	0.19	-2.54	-4.99	-2.26	2.73	2.69
4	**TM_42**	0.36	-2.35	-5.16	-2.46	2.70	2.65
5	**TM_43**	0.50	-2.26	-5.30	-2.55	2.76	2.76

The HOMO and LUMO energy levels calculated from the CV measurements provide further arguments for the smaller band gap for the increased π-systems.

The HOMO energy levels for all five synthesized tetra-substituted target materials are similar to those of the corresponding di-substituted target materials (see Table 4). However, the LUMO energy levels are stabilized by approximately 0.3 eV for each target material, thus decreasing the band gaps. The trend of HOMO energy levels is similar to the di-substituted target materials, with **TM_40** (TetraTPAThio-PCP) having the deepest HOMO energy level of the four triarylamine-substituted target materials (Table 10, entry 2). At the same time, **TM_39** and **TM_41** contain the shallowest HOMO energy levels, with -5.05 eV for **TM_39** (Tetra-*p*-OMeTPAThio-PCP, Table 10, entry 1) and -4.99 eV for **TM_41** (Tetra-*p*-OMeTPA-EDOT-PCP, Table 10, entry 3). The HOMO energy level of the *tert*-butyl-substituted target material **TM_42** again is in between the non-substituted and the methoxy-substituted target materials (Table 10, entry 4). As expected from the results in Table 4, the HOMO energy level of **TM_43** is similar to **TM_37** and the deepest of the five target materials. Nonetheless, all five target materials are in the range of typical perovskite active materials and could provide good hole transport while prohibiting electron extraction. The optical and electrochemical band gaps are complementary and support using these materials as HTMs.

Supplementary theoretical calculations of the energetic structure of the molecules were again carried out by DAVID ELSING and are compared with the experimental results in Table 11. For **TM_41,** no theoretical calculations are available, so the discussion will only include the other four target materials.

Table 11: Comparison between experimental and theoretical energy levels of tetra-substituted [2.2]para-cyclophanes.

Entry	Compound	$E_{HOMO}^{calc.}$ [eV]	E_{HOMO}^{CV} [eV]	E_{LUMO}^{calc} [eV]	E_{LUMO}^{CV} [eV]
1	TM_39	-4.58	-5.05	-1.83	-2.43
2	TM_40	-4.84	-5.19	-1.94	-2.48
3	TM_42	-4.78	-5.16	-1.92	-2.46
4	TM_43	-4.86	-5.30	-2.02	-2.55

For the calculated values in Table 11, a similar trend as in Table 5 is observable. The experimental energy levels of the HOMO are deeper, while the LUMO energy levels are stabilized. The difference is minimally larger than for the di-substituted target materials ($\sim$ 0.2 eV, see Table 5), and this small change could be explained by stronger intermolecular interactions between the four arms of the tetra-substituted [2.2]paracyclophanes compared to the two arms of the di-substituted [2.2]paracyclophanes. In any way, the trend for both the calculated and the experimental HOMO and LUMO energy levels is similar and supports the ranking of different donor and π-bridge structures from chapter 3.2.2.

In addition, the results obtained for the tetra-substituted [2.2]paracyclophanes can be compared to the results of the publications from PARK et al. and LIN et al.[163, 165] The properties of the two synthesized materials from Figure 21 are provided in Table 12, together with the results of the methoxy-substituted **TM_39, TM_41,** and **TM_43,** as the structures are closest to the published structures.

Table 12: Comparison of synthesized target materials and similar structures from literature.

Entry	Compound	λ_{max} [nm]	E_{HOMO}^{CV} [eV]	$E_{HOMO}^{calc.}$ [eV]	ΔE_g [eV]
1	TM_39	416	-5.05	-4.58	2.60
2	TM_41	410	-4.99	-	2.69
3	TM_43	379	-5.30	-4.86	2.76
4	PCP-TPA[163]	340	-5.34	-4.44	2.97
5	WS-1[165]	365	-5.25	-4.39	2.43

The comparison of the different structures provides interesting insights regarding the properties. The use of thiophene and EDOT as π-bridges leads to an increased red-shift of the absorbance maximum, with **TM_39** possessing a λ_{max} at 416 nm (Table 12, entry 1) while the corresponding literature materials absorb strongest at 340 nm and 365 nm for PCP-TPA and WS-1 respectively (Table 12, entries 4 and 5). The stronger red-shift for the target materials synthesized in this thesis can be attributed to the increase of the π-system with thiophene and EDOT in addition to the methoxy-substituted triarylamines. The ethene-bridged material WS-1 is red-shifted compared to PCP-TPA by 15 nm and further encourages this effect. The experimental HOMO energy levels provide similar changes in their values. The deepest HOMO energy level is attributed to PCP-TPA at -5.34 eV, in which the methoxy-substituted triaryl-amines are directly bound to the [2.2]paracyclophane.

Interestingly, the effect of the carbazole moiety is stronger than the difference between the ethene bridge and thiophene bridge, as the HOMO energy level of **TM_43** (Table 12, entry 3) is the second deepest at -5.30 eV. After that, the HOMO energy levels decrease from WS-1 over **TM_39** to **TM_41,** as expected. Regarding the calculated values, the difference is more substantial, with both literature values being shallower than the calculated values for **TM_39** and **TM_43**, but this can be attributed to the use of two different optimization strategies with different programs (GAUSSIAN *versus* TURBOMOLE). The optical band gaps depict the expected trend for the three composite target materials and PCP-TPA, with PCP-TPA containing the largest optical band gap of 2.97 eV and then decreasing from **TM_43** to **TM_39**. Interestingly, the experimental optical band gap of WS-1 (2.43 eV, Table 12 entry 5) is even smaller than for the newly synthesized compounds.

In conclusion, the important values from optical and electrochemical character-ization have been determined and visualized. With absorption maxima around 440 nm, all five compounds do not risk competition with the perovskites.[232] The optically and electrochemically calculated band gaps are large enough to prevent electron extraction, and the HOMO energy levels are in the desired range. Adding two more arms compared to the di-substituted [2.2]paracyclo-phanes resulted in consistent electrochemical properties and further promoted the dependence and influence of the different donors and π-bridges. The comparison with published literature data displayed similarities, while the influence of the π-bridges was observable. The applied synthetic strategy proved meaningful, and the synthesized materials were subjected to further testing, which will be discussed in the next chapter.

3.3.3 Applicability of tetra-substituted [2.2]Paracyclophanes

Similar to the di-substituted target materials, the further suitability of the synthesized materials was investigated next. In TGA, all four of the five compounds display stability to approximately 400 °C (chapter 8.1, Figure 63). Only for the EDOT-containing compound **TM_41**, the thermal decomposition started at around 325 °C.

After analyzing the TGA measurements, the target materials **TM_39**, **TM_40**, **TM_42,** and **TM_43** were delivered to ALEXANDER SCHULZ again. The following results were obtained by him and are presented after a thorough discussion to highlight the most interesting properties. The synthesis of **TM_41** was not carried out in time for a more thorough investigation.

Keeping the results from the di-substituted target materials in mind, the conventional solar cell architecture was chosen directly as the solubility of all four compounds in DMF was excellent. First up, the solubility of the target materials in chlorobenzene was investigated. The results are presented in Table 13.

Table 13: Results of the solubility tests for tetra-substituted [2.2]paracyclophanes.

Entry	Compound	Solubility in chlorobenzene [g l^{-1}]
1	TM_39	> 10
2	TM_40	> 10
3	TM_42	> 10
4	TM_43	2< TM < 10

Here an interesting change compared to the di-substituted target materials can be observed. For **TM_40** (TetraTPAThio-PCP, Table 13, entry 2), the solubility could be increased from < 10 g l^{-1} to > 10 g l^{-1} by the addition of two more arms (see Table 6, entry 2). However, the solubility for the carbazole-containing target materials **TM_43** decreased significantly compared to the di-substituted **TM_37**.

Next up, the ionization potential of the four compounds was measured *via* PESA and compared to the theoretical results of DAVID ELSING. The results are provided in Table 14.

Table 14: Comparison of the experimental and calculated ionization potentials.

Entry	Compound	Exp. IP [eV]	Calc. IP [eV]
1	TM_39	5.27	5.26
2	TM_40	5.42	5.58
3	TM_42	5.34	5.43
4	TM_43	5.43	5.55

When comparing the results from Table 14, a similar trend to the di-substituted target materials can be observed (see Table 7). The ionization potentials are shallower for the experimental values, but the difference between the experimental and calculated values is not significant. The experimental and calculated values for all four compounds are in the anticipated and necessary range for further application. For **TM_39**, a device could be manufactured and the results are provided in chapter 3.6.1.

3.4 *MESO*-TETRAKIS-SUBSTITUTED PORPHYRINS

The third building block of this work is the porphyrin. Porphyrins have been known as a widely studied macrocycle with excellent stability and useful optical and photophysical properties.[233] Apart from easy functionalization along the different positions of the macrocycle, the diversity of the porphyrin is further increased by the ability to coordinate various metals. For further discussion, an overview of the different substitution possibilities is provided in Figure 47.

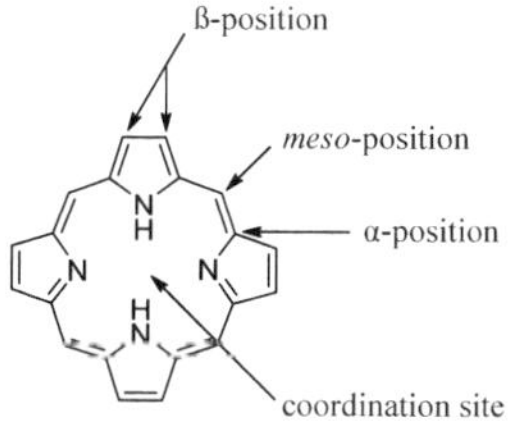

Figure 47: Nomenclature of different substitution positions in porphyrins.

For the following chapter, only the substitution at the *meso*-position will be looked into as all syntheses were performed with *meso*-tetrakis-(4-bromophenyl)porphyrin as starting material which was kindly provided by LUKAS LANGER (working group BRÄSE). Several publications in literature have described the influence of the substituents and metals on the optoelectronic properties of the porphyrins and especially the substitution with triphenylamines, and the use of first-row transition metals have provided promising results.[174, 234-236]

3.4.1 Synthesis and Characterization

The synthesis of the porphyrin target materials was divided into three different steps. In the first step, the starting material and the donor moieties were coupled *via* the beforementioned NEGISHI cross-coupling conditions (chapter 3.3.1). Due to the excess use of zinc chloride, the metal was automatically incorporated into the porphyrin coordination site. In the second step, the porphyrin was demetalated, and in the third step, two different metals were incorporated, in this case nickel and cobalt.

For the first synthesis and to optimize the reaction conditions and workup, the starting material *meso*-tetrakis-(4-bromophenyl)porphyrin and the synthesized compound **12** (*p*-OMcTPAThio) were chosen. The NEGISHI reagent was prepared as described in chapter 3.3.1 and similar to the tetra-substituted [2.2]paracyclophanes a reaction time of 18 hours at 70 °C was sufficient for adequate conversion (Scheme 15). The crude product could be purified *via* column chromatography on silica gel right after the evaporation of the solvent. The remaining starting materials and the side products could be removed using mixtures of cyclohexane and tetrahydrofuran as eluents. When most byproducts were removed, the desired target material was washed down the column with pure tetrahydrofuran. During the evaporation of said tetrahydrofuran, however, parts of the solvent degraded, and the ^{1}H NMR showed signals corresponding to residues of aldehydes and alkyl chains. These residues could be removed by dissolving the target material in dichloromethane, adding methanol, and then evaporating the dichloromethane under reduced pressure. The desired target material precipitated and could be separated *via* centrifugation.

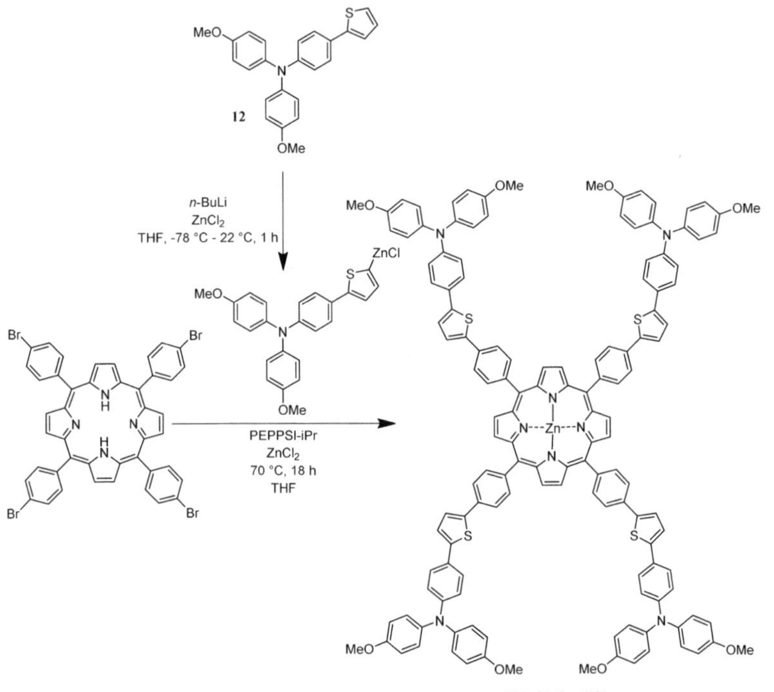

*Scheme 15: General synthesis for the NEGISHI cross-coupling exemplary for **TM_44_Zn**.*

For the coupling of the starting material and compound **12**, target material **TM_44_Zn** could be isolated as an intense red powder with a yield of 69%. Similar to the tetra-substituted [2.2]paracyclophanes, this would correspond to a conversion of 91% for each of the brominated sites. The structure and purity of **TM_44_Zn** were confirmed *via* ^{1}H NMR, ^{13}C NMR, HSQC, MS, and IR. The ^{1}H NMR of target materials **TM_44_Zn** is visualized in Figure 48. The spectrum displays the characteristic signal that was to be expected for *meso*-substituted porphyrins, as well as the previously discussed signal corresponding to the donor moieties. The signal of the eight protons from the porphyrin core can be observed at 8.97 ppm, while the signal of the 16 protons of the phenyl ring connecting the porphyrin to the thiophene is shifted to 8.23 ppm and 8.06 ppm. The signals of the overall eight protons of the thiophene units are shifted to 7.66 ppm and 7.38 ppm, and the signals of the 16 protons of the phenyl ring connecting the amine unit with the thiophene are shifted to the expected 7.56 ppm and 6.95 ppm (see Figure 41). The signals of the 32 protons of the methoxy-substituted phenyls are shifted to 7.08 ppm, and 6.88 ppm and the signal of the protons of the methoxy groups themselves is shifted to 3.77 ppm. This assignment is further supported by the ^{13}C NMR and HSQC and in ESI MS analysis, measured by and interpreted with the help of PD DR. PATRICK WEIS (IPC, KIT) for all porphyrins, the *m/z* of [M]$^+$ and [M]$^{2+}$ were found while the high resolution confirmed the exact mass of the product.

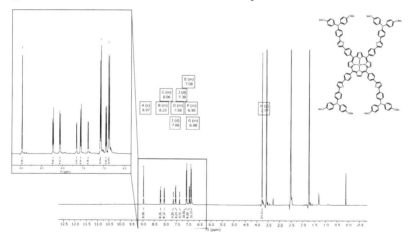

*Figure 48: ^{1}H NMR of **TM_44_Zn**, measured in THF-d8.*

After the successful synthesis of **TM_44_Zn** was confirmed and the workup procedure was optimized, further derivatives containing different donor-π-moieties were synthesized. After analysis of the results obtained for the different [2.2]paracyclophanes, only donor-π-moieties containing methoxy groups were chosen, namely the *para*-methoxy-substituted triarylamine and the methoxy-substituted carbazole. The methoxy-substituted donor-π-moieties have provided the combined best results for solubilities and the energy level of the HOMO. For both donor groups, the thiophene and the EDOT have been successfully added as π-bridges, and to further investigate the effect of the π-bridge, the OMe-TPA-selenophene donor-π-moiety **30** was introduced to the porphyrin as well. The results of the NEGISHI cross-coupling reactions are provided in Table 15.

Table 15: Overview of synthesized target materials containing zinc as metal.

Entry	Starting material	Target Material	Yield [%]
1	12	Tetra-*p*-OMeTPAThio-Por-Zn (**TM_44_Zn**)	69
2	14	Tetra-*p*-OMeTPA-EDOT-PCP-Zn (**TM_45_Zn**)	72
3	25	TetraOMeCbzThio-PCP-Zn (**TM_46_Zn**)	86
4	29	TetraOMeCbzTPA-EDOT-PCP-Zn (**TM_47_Zn**)	31
5	30	Tetra-*p*-OMeTPA-Se-PCP-Zn (**TM_48_Zn**)	35

While the high yields of both **TM_45_Zn** and **TM_46_Zn** (72% and 86% respectively, Table 15, entries 2 and 3) can be attributed to the solubility, which was sufficient in pure tetrahydrofuran and insufficient in any mixtures containing 30% or more cyclohexane in tetrahydrofuran, the low yields of **TM_47_Zn** and **TM_48_Zn** are mostly contributed to their different solubilities. For **TM_47_Zn,** the solubility in mixtures of cyclohexane and tetra-hydrofuran was too good, resulting in additional workup steps and thus decreasing the isolated yield (31%, Table 15, entry 4), the solubility of **TM_48_Zn** in pure tetrahydrofuran was dissatisfactory. Pyridine had to be added during column chromatography to allow isolation of the target material and significantly lowering the yield (35%, Table 15, entry 5). All five target materials were isolated as dark solids with intense red fluorescence.

After ruling out **TM_48_Zn** for any further modification due to the low solubility, the next step was to demetalate the synthesized target materials. For this, the typical demetallation reagent trifluoroacetic acid was used.[237] The zinc-containing materials **TM_44_Zn,** **TM_45_Zn,** **TM_46_Zn,** and **TM_47_Zn** were dissolved in dichloromethane, and an excess of trifluoroacetic acid was added (Scheme 16). The reaction mixture was stirred for 1.5 hours, and after neutralizing and washing the product mixture, the desired demetalated target materials were obtained by precipitation from a dichloro-methane/cyclohexane mixture.

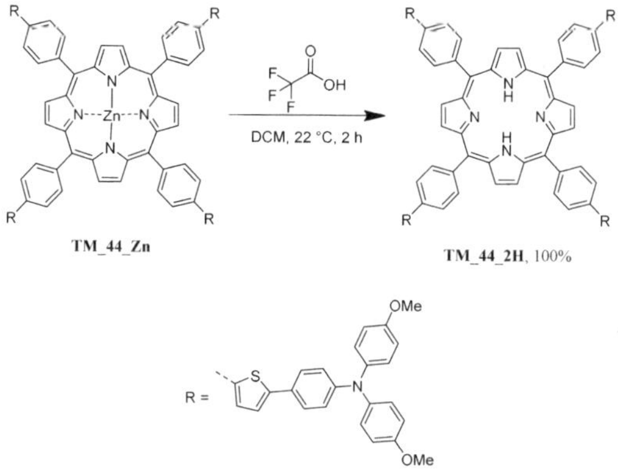

Scheme 16: General synthesis for the demetallation exemplary for TM_41_2H.

The demetallation was quantitative for all four compounds, resulting in the successful synthesis of the target materials **TM_44_2H, TM_45_2H, TM_46_2H,** and **TM_47_2H** from the corresponding zinc porphyrins. The structure and purity of the synthesized metal-free porphyrins were again confirmed by ¹H NMR, ¹³C NMR, HSQC, MS, and IR. The ¹H NMR of target material **TM_44_2H** is visualized in Figure 49 to illustrate the differences compared to the zinc-containing target material **TM_44_Zn** (Figure 48). The signal of the protons from the N-H of the pyrrole subunits is shifted to the expected -2.62 ppm, confirming the successful demetallation.[238] The signal contributed to the N-H hydrogens could be observed for all four target materials, while no residual signals from the zinc-containing porphyrin were observed in any of the ¹H NMR spectra.

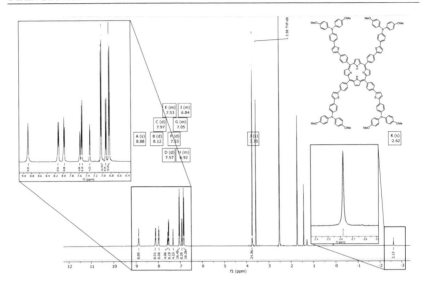

*Figure 49: ^1H NMR of **TM_44_2H**, measured in THF-d8.*

The results of the demetallation are summarized in Table 16.

Table 16: Overview of synthesized target materials with free coordination site.

Entry	Starting material	Target Material	Yield [%]
1	TM_44_Zn	Tetra-*p*-OMeTPAThio-Por-2H (**TM_44_2H**)	100
2	TM_45_Zn	Tetra-*p*-OMeTPA-EDOT-PCP-2H (**TM_45_2H**)	100
3	TM_46_Zn	TetraOMeCbzThio-PCP-2H (**TM_46_2H**)	100
4	TM_47_Zn	TetraOMeCbzTPA-EDOT-PCP-2H (**TM_47_2H**)	99

Due to the effortless and straightforward demetallation procedure, further modifications could be realized. With literature reporting promising optoelectronic results for different first-row transition metals, nickel and cobalt were chosen as new metal centers inside the porphyrin moiety.[179] For this synthesis, the standard literature procedures were applied.[239] The metal-free porphyrin was dissolved in *N,N*-dimethylformamide, and excess nickel acetate tetrahydrate or cobalt acetate tetrahydrate were added, respectively. The reaction mixture was heated to 140 °C for 22 hours, and the crude product was purified *via* column chromatography on silica gel after evaporation of the solvent to remove tetrahydrofuran residues; the product was precipitated from a dichloromethane/cyclohexane mixture (Scheme 17).

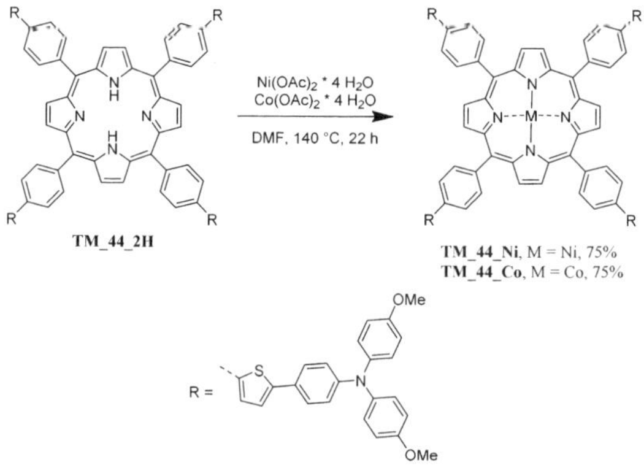

Scheme 17: *General syntheses for the metalation with nickel and cobalt, exemplary for* **TM_44_Ni** *and* **TM_44_Co**.

Both reactions were successful for each of the four demetalated porphyrins, presenting eight new porphyrin target materials overall. An overview of the different target materials and their yields is provided in Table 17.

Table 17: *Overview of the synthesized target materials containing nickel or cobalt as metals.*

Entry	Starting material	Metal salt	Target Material	Yield [%]
1	TM_44_2H	Ni(OAc)$_2$*4 H$_2$O	TM_44_Ni	75
2	TM_45_2H	Ni(OAc)$_2$*4 H$_2$O	TM_45_Ni	74
3	TM_46_2H	Ni(OAc)$_2$*4 H$_2$O	TM_46_Ni	87
4	TM_47_2H	Ni(OAc)$_2$*4 H$_2$O	TM_47_Ni	64
5	TM_44_2H	Co(OAc)$_2$*4 H$_2$O	TM_44_Co	75
6	TM_45_2H	Co(OAc)$_2$*4 H$_2$O	TM_45_Co	76
7	TM_46_2H	Co(OAc)$_2$*4 H$_2$O	TM_46_Co	78
8	TM_47_2H	Co(OAc)$_2$*4 H$_2$O	TM_47_Co	49

The yields for the metalation reaction are all in a similar range, mostly between 74% and 78%. While **TM_46_Ni** stands out with a higher yield of 87%, both target materials containing the OMeCbz-EDOT scaffold (**TM_47_Ni** and **TM_47_Co**) could only be isolated with lower yields (64% and 49%, respectively). For all eight compounds, the losses can be attributed to the workup as all compounds dissolved better in the cyclohexane/dichloromethane mixture than the zinc and metal porphyrins, with the best solubility for the **TM_47** scaffold. All eight compounds were investigated regarding their structure and purity by ^{1}H NMR, ^{13}C NMR, MS, IR, and, if possible, HSQC. The ^{1}H NMR of **TM_44_Ni** is displayed in Figure 50.

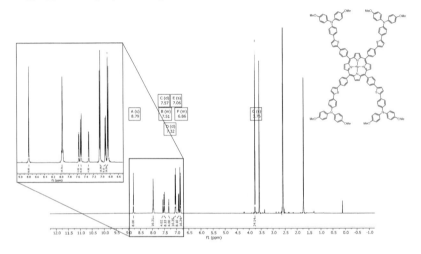

*Figure 50: ^{1}H NMR of **TM_44_Ni**, measured in THF-d8.*

Similar to the ^{1}H NMR spectra of **TM_44_Zn** and **TM_44_2H**, the shift of the signals contributed to the protons of the donor groups is minimal, and all expected signals are observable. To prove the metalation, the area from -0.6 ppm to -3.2 ppm displays no signal corresponding to the N-H hydrogens, confirming that the desired target material was indeed obtained. While the NMR spectra of **TM_44_Ni** and the further synthesized nickel-containing porphyrins were easily interpretable and provided accurate resolutions, the NMR analysis of the cobalt-containing porphyrins proved more difficult due to the paramagnetic nature of Co^{2+}. For the cobalt-containing porphyrins, the detected signals in the ^{1}H NMR displayed the expected shifts, but integration was not sufficiently possible due to the broadness of the signals. With the help of ESI MS and the corresponding high-resolution of the obtained signals, the mass and, therefore, the target materials' structure could be confirmed.

In conclusion, an overall number of 17 new porphyrins could be synthesized. The structure and purity were thoroughly investigated and could be confirmed for all target materials. While the yield for the NEGISHI cross-coupling reaction and the metalation reaction was strongly influenced by the solubility, the demetallation provided the desired target materials with quantitative yield. **TM_48_Zn** was not further modified due to the extremely low solubility. The next chapter will investigate all 17 synthesized target materials regarding their optoelectronic properties.

3.4.2 Optical and Electrochemical Characterization

After the successful synthesis of the 17 porphyrins could be confirmed, the next step was the analysis of the optical and electrochemical properties. To allow an easier overview, the target materials will be separated into four groups: the zinc-containing target materials, the metal-free target materials, the nickel-containing target materials, and the cobalt-containing target materials. The groups will be discussed in the order of the synthesis, and thus, the UV-vis and photoluminescence spectra of the zinc-containing target materials are visualized in Figure 51.

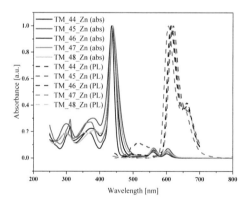

Figure 51: Normalized UV-vis spectra (in THF, lines) and normalized photoluminescence spectra (in THF, strokes) of the zinc-containing porphyrins.

The qualitative UV-vis measurements (Figure 51 abs) were performed in THF, and the maximum absorbance wavelength was used for the photoluminescence measurements (Figure 51 PL). When comparing the UV-vis spectrum of the zinc-containing porphyrins (Figure 51) to the UV-vis spectra of the [2.2]-para-cyclophanes (Figure 34, Figure 35, and Figure 44), the presence of three other bands for the porphyrins is observed. The strongest blue-shifted band at 300 nm for **TM_44_Zn**, **TM_45_Zn**, and **TM_48_Zn** and at 310 nm for the carbazole-containing target materials **TM_46_Zn** and **TM_47_Zn** can again be contributed to a $\pi \rightarrow \pi^*$ transition localized in the donor moiety.[221] The second band, which was also observable for the beforementioned [2.2]paracyclophane compounds, can be observed between 360 nm and 385 nm. Similar to the [2.2]paracyclophanes, this transition could be attributed to an intramolecular charge transfer between the donor and the π-bridges.[165] For this absorbance, an increase in electron density in the π-bridge leads to a red shift of the absorbance from 373 nm for **TM_44_Zn** to 380 nm for **TM_46_Zn** and from 361 nm for **TM_45_Zn** to 372 nm for **TM_47_Zn**. The absorbance for the selenophene-containing **TM_48_Zn** displays the strongest red shift at 385 nm. While these absorbance wavelengths provide additional information on the donor groups and can be interpreted as a shortage of electronic interaction between donors and core, the more interesting bands are the characteristic bands of the porphyrin structure.[240] These bands can be divided into two distinct sets of electronic transitions. The bands corresponding to the transition from the ground state to the first excited state ($S_0 \rightarrow S_1$) are called the Q bands between 500 nm and 700 nm. The blue-shifted band between 380 nm and 500 nm corresponds to the transition from the ground state to the second excited state ($S_0 \rightarrow S_2$) and is called the Soret band.[241] Only minor differences in absorbance wavelength are observable for the Soret band of the zinc-containing porphyrins. For **TM_46_Zn,** the band is blue shifted the most at 435 nm, while **TM_47_Zn** and **TM_48_Zn** display the largest absorbance wavelength at 439 nm, which is the typical range for zinc-containing porphyrins.[234] The trend regarding the absorbance is similar to the bands at around 370 nm, but the smaller shift shows the minor influence of the donor moieties. For the zinc-coordinated porphyrins, two Q bands are observable. The first Q band is located at 560 nm, and the second at 602 nm. Here the differences between the five target materials are insignificant. All five zinc-containing porphyrins display photoluminescence between

604 nm and 622 nm with a weak red-shifted shoulder. Here **TM_47_Zn** is blue shifted the furthest, and from **TM_46_Zn**, **TM_44_Zn**, and **TM_48_Zn,** the strongest red shift is observed for **TM_45_Zn**. The strong photoluminescence of the zinc-containing porphyrins can be attributed to the full valence electron orbital (d^{10}) of the Zn^{2+} ion.[242] The results of the optical analysis are displayed in Table 18.

After analyzing the optical properties of the zinc-containing porphyrins, the next series of target materials are the metal-free porphyrins. The qualitative UV-vis and photoluminescence measurements are visualized in Figure 52. Similar to the zinc-containing porphyrins, the bands corresponding to the donor-π-moieties are observable at around 305 nm and 370 nm. However, the Soret band of the metal-free porphyrins is blue shifted between 7 nm and 15 nm, and the triarylamine-containing **TM_44_2H** and **TM_45_2H** target materials are blue shifted more strongly than **TM_46_2H**. In addition to the blue shifted Soret bands, the metal-free porphyrins display four different Q bands. These four Q bands can be explained by the four-orbital-model theory, in which the asymmetric structure of the metal-free porphyrin leads to a split of the degenerated orbital energy levels.[241, 243] The Q bands for all four target materials are similar and are displayed from 520 nm, over 562nm, and 599 nm to 653 nm.

Regarding their photoluminescence, all four target materials show a band between 659 nm and 664 nm. Here the thiophene-containing target materials **TM_44_2H** and **TM_46_2H** show a slight blue shift compared to the EDOT-containing target materials. The results of the optical analysis are displayed in Table 18.

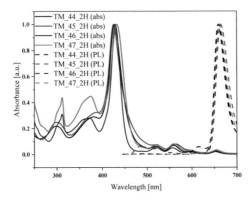

Figure 52: Normalized UV-vis spectra (in THF, lines) and normalized photoluminescence spectra (in THF, strokes) of the metal-free porphyrins.

The UV-vis spectra of the nickel- and cobalt-containing porphyrins show the same expected bands for the donor-π-moieties between 300 nm and 380 nm (Figure 53). The difference between the Soret bands for the nickel-containing porphyrins (426 nm to 433 nm) and the metal-free porphyrins is minimal, while for the cobalt-containing porphyrins, the Soret bands between 431 nm and 438 nm are more similar to the ones of the zinc-containing porphyrins. The nickel-containing porphyrins' Q bands are located around 533 nm and 579 nm, while the Q bands of the cobalt-containing porphyrins are located around 545 nm and 594 nm. Similar to the zinc-containing porphyrins, the other metal-containing porphyrins are symmetric again and only display two Q bands. The results of the optical analysis are displayed in Table 18.

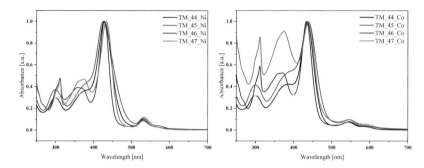

Figure 53: Left: Normalized UV-vis spectra (in THF) of the nickel-containing porphyrins. Right: Normalized UV-vis spectra (in THF) of the cobalt-containing porphyrins.

No photoluminescence could be observed or recorded for nickel- and cobalt-containing porphyrins. Similar effects are described in the literature and could be explained by the paramagnetic nature of the metal centers.[244] This paramagnetic property of Ni^{2+} and Co^{2+} could lead to fluorescence quenching by enhanced spin-orbit coupling and non-radiative deactivation processes.[179]

As no photoluminescence was observable for the nickel- and cobalt-containing compounds and the Q bands and PL spectra overlayed for the zinc-containing and metal-free porphyrins, the optical band gap could not be determined by Equation 3. However, the optical band gap can also be calculated from the absorbance onset wavelength:

$$\Delta E_g = \frac{1243}{\lambda_{onset}} \; eV \qquad (5)$$

Equation 5: Alternative calculation of the optical band gap.[179]

An overview of all optical properties of the porphyrin target materials is given in Table 18.

Table 18: Overview of optical properties of the porphyrin target materials.

Entry	Compound	$\lambda_{abs\ donor}$ [nm]	Soret band [nm]	Q bands [nm]	λ_{onset} [nm]	ΔE_g [eV]
1	TM_44_Zn	299, 374	436	561, 603	624	1.99
2	TM_45_Zn	300, 379	438	561, 603	629	1.97
3	TM_46_Zn	311, 363	435	561, 603	621	2.00
4	TM_47_Zn	311, 367	439	561, 603	626	1.98
5	TM_48_Zn	298, 379	439	561, 603	628	1.97
6	TM_44_2H	298, 378	427	520, 559, 597, 653	671	1.85
7	TM_45_2H	299, 384	424	520, 564, 600, 655	679	1.83
8	TM_46_2H	311, 372	429	520, 556, 595, 654	673	1.84
9	TM_47_2H	311, 374	432	520, 560, 598, 654	679	1.83
10	TM_44_Ni	298, 375	427	533, 574	606	2.05
11	TM_45_Ni	299, 375	424	533, 574	611	2.03
12	TM_46_Ni	311, 361	428	533, 574	620	2.00
13	TM_47_Ni	311, 373	432	533, 574	608	2.04
14	TM_44_Co	298, 379	435	543, 590	627	1.98
15	TM_45_Co	300, 385	438	545, 590	638	1.94
16	TM_46_Co	311, 370	433	545, 590	652	1.90
17	TM_47_Co	311, 374	438	546, 592	646	1.92

The optical band gaps of all porphyrin materials are smaller than for the [2.2]paracyclophanes, and while the donor-π-moieties only have an insignificant influence on the optical band gap, the influence of the coordination site is stronger. This effect was expected as the difference between HOMO and the macrocycle mostly determines the LUMO energy levels of the porphyrin.[175] The smallest band gap was obtained for the metal-free porphyrins (1.83 eV – 1.85 eV, Table 18, entries 6, 7, 8, and 9), while the different metals all contain similar band gaps between 1.90 eV and 2.05 eV. Nonetheless, an optical band gap of around 2.00 eV is large enough to prevent electron and hole extraction.[175]

Next, cyclic voltammetry was conducted to determine the synthesized porphyrins' experimental HOMO and LUMO energy levels. The conditions can be found in chapter 3.2.2. The obtained cyclic voltammograms are divided into the oxidative and the reductive cycles and are set up in the same groups as in the optical measurements. The cyclic voltammograms are visualized in Figure 54.

For the zinc-containing porphyrins, the first oxidation process can be observed between 0.1 V and 0.7 V (Figure 54, row 1, right). For the target materials **TM_45_Zn, TM_47_Zn,** and **TM_48_Zn,** the oxidation process is quasi-reversible, while the first oxidation process for the two thiophene-containing target materials **TM_44_Zn** and **TM_46_Zn** are irreversible. The selenophene-containing **TM_48_Zn** displays a second oxidation around 0.2 V after the first oxidation, and when comparing the donor moieties, the first oxidation of the triarylamine-containing target materials is easier to achieve than the first oxidation of the carbazoles. Regarding the corresponding reductive cyclic voltammograms (Figure 54, row 1, left), all five target materials display multiple reductive processes between -1.2 V and -3.0 V. While similar quasi-reversible reductive processes between -2.5 V and -3.0 V were also observed for the [2.2]para-cyclophanes and could be contributed in the donors, for the porphyrins the first two reductive processes between -1.2 V and -2.5 V could be contributed to the porphyrin core. Here, four of the five target materials display a first reversible reduction process around -1.8 V, while **TM_44_Zn** displays its first quasi-reversible reduction process earlier at around -1.3 V. The resulting energy levels are provided in Table 19.

For the oxidative cyclic voltammograms of the metal-free porphyrins (Figure 54, row 2, right), again, one oxidation process is observable in the range from 0.1 V to 0.6 V. Here, the triarylamine-containing target materials **TM_44_2H** and **TM_45_2H** display reversible oxidation processes from 0.1 V to 0.3 V while the carbazole-containing **TM_46_2H** and **TM_47_2H** display irreversible oxidation processes at higher potentials (0.4 V to 0.6 V). The reductive cyclic voltammograms (Figure 54, row 2, left) of all four materials display a first reduction process around -1.5 V. While the reduction processes for **TM_45_2H** and **TM_46_2H** are reversible, the first reduction processes of **TM_44_2H** and **TM_47_2H** are much less distinct and of quasi-reversible nature. Apart from **TM_46_2H**, the other three target materials display a second reduction at around -2.0 V. Similar to the beforementioned zinc-containing porphyrins, different reductive processes are observable in the range from -2.5 V to -3.0 V, which again could be contributed to reduction processes on the donor-π-moieties. The resulting HOMO and LUMO energy levels are provided in Table 19.

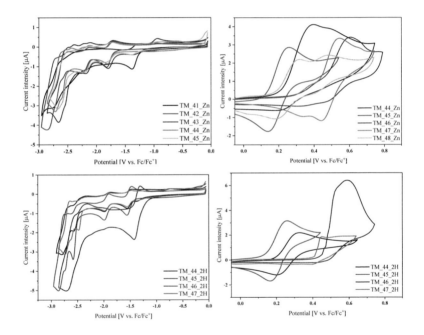

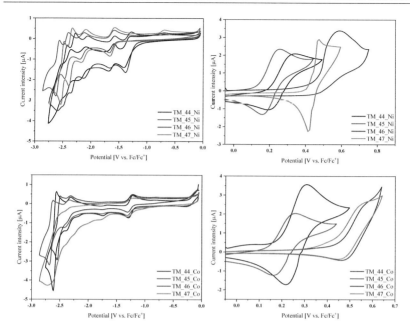

Figure 54: Cyclic voltammograms (left: reductive, right: oxidative) of the synthesized porphyrins. First row: Zinc; Second row: Metal-free; Third row: Nickel; Fourth row: Cobalt.

The oxidative cyclic voltammograms of the nickel- and the cobalt-containing porphyrins illustrate a similar trend as already seen for the previously mentioned metal-free and zinc-containing porphyrins (Figure 54, rows 3 (Ni) and 4 (Co), right). The triarylamine-containing target materials display a first oxidation process between 0.1 V and 0.3 V as a reversible process. In comparison, the carbazole-containing target materials display their first oxidation processes at higher potentials between 0.4 V and 0.7 V. The first oxidative processes of the cobalt-containing porphyrins **TM_46_Co** and **TM_47_Co** are quasi-reversible, and the nickel-containing porphyrins **TM_46_Ni** and **TM_47_Ni** display the first oxidative processes that are reversible and irreversible, respectively. The reductive cyclic voltammograms of the nickel-containing porphyrins (Figure 54, row 3, left) display a first reductive process for the thiophene-containing **TM_44_Ni** and **TM_46_Ni** at around -1.4 V. In contrast, the EDOT-containing **TM_45_Ni** and **TM_47_Ni** display their first reductive process around -1.6 V. In comparison to this, the reductive cyclic voltammograms of the cobalt-containing

103

porphyrins (Figure 54, row 4, left) all show a first reductive process at the same potential (-1.3 V). For both metals, the expected further reductive processes of the donors can be observed between -2.5 V and -3.0 V. The results of all electrochemical analyses are provided in Table 19.

Table 19: Electrochemical properties of all synthesized porphyrins.

Entry	Compound	E_{ox} [V vs. Fc]	E_{red} [V vs. Fc]	E_{HOMO} [eV]	E_{LUMO} [eV]	$E_{HOMO-LUMO}$ [eV]	ΔE_g [eV]
1	TM_44_Zn	0.30	-1.34	-5.10	-3.46	1.64	1.99
2	TM_45_Zn	0.20	-1.77	-5.00	-3.03	1.97	1.97
3	TM_46_Zn	0.56	-1.76	-5.36	-3.05	2.31	2.00
4	TM_47_Zn	0.43	-1.77	-5.23	-3.04	2.19	1.98
5	TM_48_Zn	0.18	-1.33	-4.98	-3.48	1.50	1.97
6	TM_44_2H	0.26	-1.51	-5.06	-3.30	1.76	1.85
7	TM_45_2H	0.21	-1.52	-5.01	-3.29	1.72	1.83
8	TM_46_2H	0.48	-1.36	-5.28	-3.44	1.84	1.84
9	TM_47_2H	0.43	-1.52	-5.23	-3.28	1.95	1.83
10	TM_44_Ni	0.26	-1.34	-5.06	-3.46	1.60	2.05
11	TM_45_Ni	0.21	-1.64	-5.01	-3.29	1.72	2.03
12	TM_46_Ni	0.55	-1.35	-5.35	-3.46	1.89	2.00
13	TM_47_Ni	0.44	-1.62	-5.24	-3.18	2.06	2.04
14	TM_44_Co	0.26	-1.25	-5.06	-3.56	1.50	1.98
15	TM_45_Co	0.21	-1.28	-5.01	-3.52	1.49	1.94
16	TM_46_Co	0.51	-1.27	-5.31	-3.53	1.78	1.90
17	TM_47_Co	0.50	-1.27	-5.30	-3.54	1.76	1.92

The results of the experimental analysis of the HOMO and LUMO energy levels provide more interesting insights into the influence of the different building blocks. While the optical band gap is similar between all metal-containing porphyrins, the band gap obtained through the HOMO and LUMO energy levels is substantially different between the metals and the donor groups. Larger band gaps are achieved by substitution with carbazoles (Table 19, entries 3, 4, 8, 9, 12, 13, 16, and 17) with approximately 0.2 eV difference to the triarylamine-containing target materials. A similar trend regarding the donor groups is observable for the HOMO energy level. While the donor is comparably shallow with the triarylamines at around -5.0 eV, the substitution with carbazole groups leads to a deeper HOMO at around -5.3 eV. The component in the coordination site does not seem to influence the HOMO energy at all. However, the LUMO energy levels are more strongly influenced by the different metals or metal-free centers. While the deviation for the HOMO energy level of the **TM_44** for the different centers is only 0.04 eV (Table 19, entries 1, 6, 10, and 14), the LUMO energy levels deviate by 0.26 eV. Further comparison with theoretical results is impossible as no results are yet obtained.

To summarize the optical and electrochemical characterization of the synthesized 17 porphyrins, the UV-vis spectra provided insights into whether the porphyrin would compete with the perovskite for photons. However, the Soret band is blue-shifted far enough for all materials to allow enough absorbance for the perovskite. The optical band gap and the calculated difference between the HOMO and the LUMO energy levels are large enough to guarantee the block of excited electrons and prohibit charge recombination. The HOMO energy levels of the porphyrins between -5.0 eV and -5.4 eV are in the expected range and fit well with the VB energy of the perovskites.[48] The same conclusion goes for the LUMO energy levels, which are no lower than -3.54 eV, and thus the gap between the CB energy level of the perovskite and LUMO of the porphyrins is adequate to prohibit electron extraction.[48] Due to the well-fitting properties, all 17 target materials will be subjected to further analysis to evaluate the most promising materials better.

3.4.3 Applicability of tetra-substituted Porphyrins

With the obtained results from the optical and electrochemical characterization, all 17 target materials were tested for thermal stability (chapter 8.1, Figure 64). Except for **TM_47_Co**, all target materials displayed thermal stability between 340 °C and 450 °C. The 5 % weight loss at lower temperatures (195 °C) of **TM_47_Co** could be due to measurement fluctuation as less target material could be used for the experiment. After measuring the thermal stability, the target materials were delivered to ALEXANDER SCHULZ. He obtained all the following results, which are presented after a thorough discussion highlighting the most important properties.

First up, the solubility of the compounds in chlorobenzene was the subject of investigation. Expectedly, the selenophene-containing **TM_48_Zn** was below the anticipated 2 g L^{-1}, but for all 16 other compounds, high solubilities of more than 10 g L^{-1} were observed. Next, the ionization potential of the spin-coated thin films was measured *via* PESA. Due to the low solubility of **TM_48_Zn**, spin-coating was impossible, and no results could be obtained. The obtained results are presented in Table 20.

Table 20: Overview of the experimental ionization potentials.

Entry	Compounds	Exp. IP [eV]
1	TM_44_Zn	5.04
2	TM_45_Zn	5.32
3	TM_46_Zn	5.45
4	TM_47_Zn	5.44
5	TM_44_2H	5.08
6	TM_45_2H	5.23
7	TM_46_2H	5.48
8	TM_47_2H	5.48
9	TM_44_Ni	5.29
10	TM_45_Ni	5.20
11	TM_46_Ni	5.53
12	TM_47_Ni	5.61
13	TM_44_Co	5.37
14	TM_45_Co	5.21
15	TM_46_Co	5.51
16	TM_47_Co	5.62

In the measured ionization potentials, the effects of the metals and donor groups become even more visible. When taking **TM_44** as the first example (Table 20, entries 1, 5, 9, and 13), the difference in IP is insignificant between the zinc-containing **TM_44_Zn** (5.04 eV) and **TM_44_2H** (5.08 eV) while the IP is deeper for **TM_44_Ni** (5.29 eV) and even deeper for **TM_44_Co** (5.37 eV), which means a difference of 0.33 eV between the shallowest and the deepest potential. In comparison to these values, the ionization potentials for **TM_45** are much closer together. Here, **TM_45_Ni** has the shallowest IP with 5.20 eV, and **TM_45_Zn** contains the deepest IP with 5.32 eV, leading to a difference of only 0.12 eV. Compared to the triarylamine-containing target materials **TM_44** and **TM_45**, the carbazole-containing target materials **TM_46** and **TM_47** possess deeper ionization potentials. For **TM_46**, the ionization potentials differ from **TM_46_Zn** with 5.45 eV to **TM_46_Ni** with 5.53 eV by only 0.08 eV. A slightly larger discrepancy is measured for **TM_47**. Here, **TM_47_Zn** possesses the shallowest IP with 5.44 eV, while the deepest IP belongs to **TM_47_Co** with 5.62 eV, leading to a difference of 0.18 eV. While **TM_44, TM_46,** and **TM_47** all have similar trends regarding their ionization potentials, with the zinc-containing porphyrin having the shallowest and the cobalt- and nickel-containing porphyrins having similar deeper ionization potentials, **TM_45** does not comply with the trend. Here the deepest ionization potential is achieved with the zinc-containing porphyrin, while the nickel- and cobalt-containing porphyrins contain the shallowest ionization potentials. As the IP was measured in thin films, compared to the HOMO and LUMO energy levels, which were measured in solution, the stacking and order in the thin films might lead to stronger or weaker intermolecular interaction, thus influencing the ionization potentials stronger than when dissolved.

The application as hole transport materials is ongoing, so that no results can be presented yet. To summarize the current results, the investigation of the solubilities has shown that except for the selenophene-containing compound **TM_48_Zn,** all porphyrins could be used in spin-coating. The PESA measurements' results have confirmed the cyclic voltammetry findings. All zinc-containing porphyrins and the cobalt- and nickel-containing target materials **TM_44** and **TM_45** contain fitting energy levels, while the energy levels of the cobalt- and nickel-containing target materials **TM_46** and **TM_47** might be too deep for efficient hole extraction.

3.5 TETRA-*PARA*-SUBSTITUTED TETRAPHENYLETHENES

After investigating the effects of different substituents at the donor groups, the influence of the π-bridges, and the different number of arms on the [2.2]para-cyclophane and further deepening the investigation by implementing the porphyrin as core which additionally allowed the variation of the coordination site, one last effect should be investigated, the influence of different methoxy-substituents on the overall molecule. To prevent any unwanted effect of the beforementioned groups, the tetraphenylethene core was chosen for this investigation. The tetraphenylethene (TPE) core has already been described in the literature as a hole transport material, and by coupling the triarylamines directly to the core, some changes in properties should be observable.

3.5.1 Synthesis and Characterization

OLIVER LAMPERT carried out the following syntheses under my supervision while the characterization was done independently. For the first step, the commercially available TPE core was brominated under literature conditions published by MATT et al. (Scheme 18).[245]

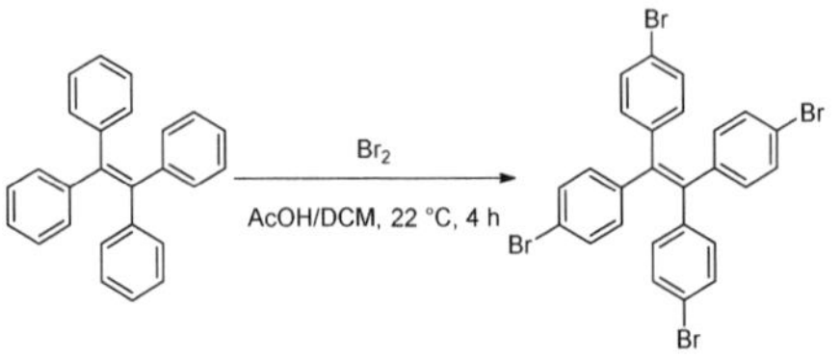

Scheme 18: Synthesis of Br$_4$-TPE.

The desired compound **49** could be isolated as a colorless solid with an 87% yield. In the second step, the corresponding coupling partners were synthesized from the triarylamines **1, 2, 3, 4,** and **5**. These triarylamines were chosen as **1** was methoxy-substituted in the *para*-position, **2** in the *meta*-position, **3** in the *ortho*-position, **4** in *para*- and one *ortho*-position, and **5** in the *para*- and both *meta*-positions. Without the issue of borylating a thiophene, typical MIYAURA borylation conditions could be applied, following the proposed mechanism from Scheme 9a.[209] The starting material **1** was dissolved in 1,4-dioxane with Pd(dppf)Cl$_2$, potassium acetate, and bis-pinacolato-diboron (B$_2$pin$_2$). After a reaction time of 18 hours at 100 °C, the desired target material 4-pinacolboryl-*N,N*-bis(4-methoxyphenyl)aniline **50** could be isolated after purification *via* column chromatography with a yield of 54% (Scheme 19).

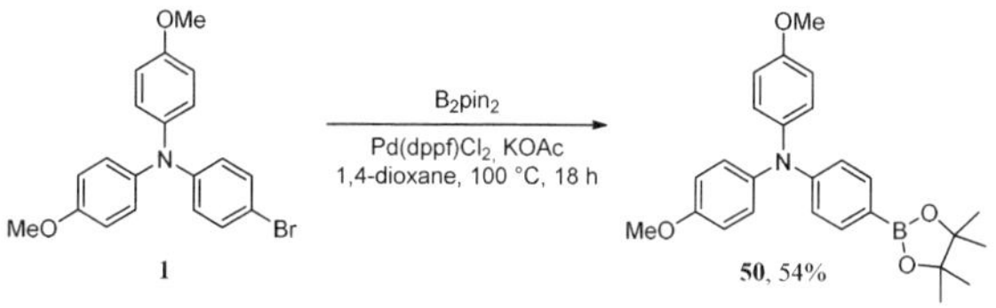

Scheme 19: Synthesis of the borylated triarylamines with compound **50** as an example.

The borylation could be repeated for the other four compounds, and an overview of the obtained yields is given in Figure 55. For starting material **2**, the corresponding borylated triarylamine **51** could be isolated with a yield of 87%, similar to compound **52**. While the yield of the borylated compound **53** from starting material **4** was similar to the synthesis of compound **50**, compound **54** could be isolated in a nearly quantitative yield (97%).

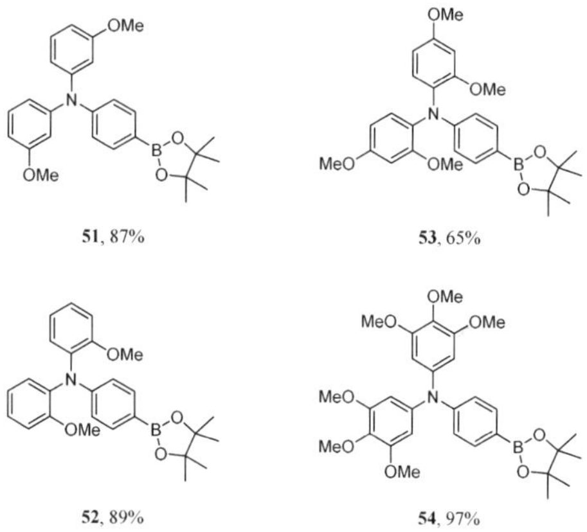

Figure 55: Overview of synthesized borylated triarylamines.

Following the successful syntheses of the desired coupling partners for **49**, the synthesis of the target materials followed the typical beforementioned conditions of a SUZUKI-MIYAURA cross-coupling reaction. The two starting materials, **49** and the desired coupling partner **50** were dissolved in degassed dimethyl sulfoxide, with Pd(dppf)Cl$_2$ as the catalyst and potassium carbonate as the base. The reaction follows the proposed mechanism depicted in Scheme 9b, and after heating to 90 °C for 18 hours, the desired target material **TM_55** (*p*OMeTPA-TPE) could be isolated with a yield of 37% (Scheme 20).

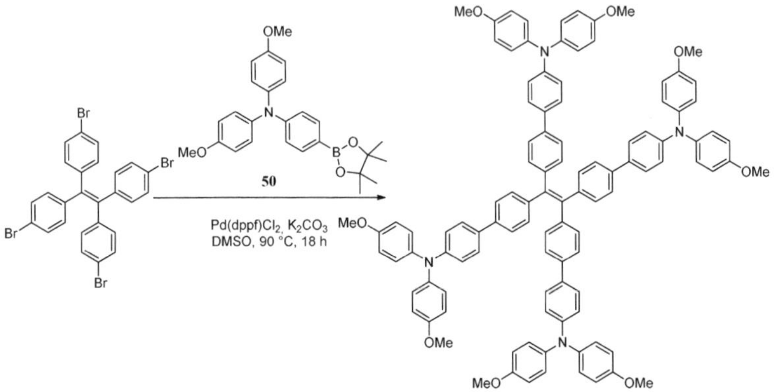

*Scheme 20: Synthesis of TPE-based target materials with **TM_55** as an example.*

The yield of the target materials largely depended on the workup's efficiency. Especially for **TM_55**, the column chromatography was strenuous, with several bands contributing to the di- and tri-coupled product with similar R_f values. An overview of the synthesized target materials is provided in Table 21.

Table 21: Overview of the synthesized target materials.

Entry	Coupling partner	Target material	Yield [%]
1	**50** (*p*OMeTPA)	**TM_55** (*p*OMeTPA-TPE)	37
2	**51** (*m*OMeTPA)	**TM_56** (*m*OMeTPA-TPE)	62
3	**52** (*o*OMeTPA)	**TM_57** (*o*OMeTPA-TPE)	71
4	**53** (*op*OMeTPA)	**TM_58** (*op*OMeTPA-TPE)	43
5	**54** (*mmp*OMeTPA)	**TM_59** (*mmp*OMeTPA-TPE)	44

The desired target materials could be isolated with sufficient yields and were obtained as bright yellow solids. All five materials were thoroughly characterized *via* [^1]H NMR, [^13]C NMR, HSQC, MS, and IR. The [^1]H NMR of **TM_55** is depicted in Figure 56. All proton signals are observable with the expected shifts. The signals from the 32 protons of the four biphenyls connecting the ethene core and the amine groups are shifted to 7.38 ppm, 7.32 ppm, 7.12 ppm, and 6.96 ppm. The signals from the 32 protons of the phenyls connecting the amine with the methoxy groups are shifted to 7.07 ppm and 6.83 ppm, and the 24 protons of the methoxy groups themselves are shifted to 3.80 ppm. Additionally, in the ^{13}C NMR, the signal of the two carbons of the ethene is observable and shifted to 140.3 ppm. High-resolution ESI MS could further confirm the correct structure.

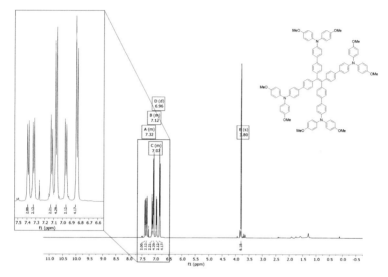

*Figure 56: ^{1}H NMR of **TM_55** in CDCl$_3$.*

In conclusion, five different target materials with varying methoxy-substitution patterns were successfully synthesized. The structure and purity were confirmed for all five target materials *via* NMR, MS, and IR. Except for **TM_55**, the target materials have not yet been described in the literature.[169] The influence of the different substitution patterns will be further discussed in the next chapter.

3.5.2 Optical and Electrochemical Characterization

After the necessary characterization and structure confirmation were concluded, the optical and electrochemical properties of the five target materials were investigated. First, UV-vis and photoluminescence were recorded, and the obtained spectra were visualized in Figure 57. The qualitative UV-vis measurements (Figure 57 abs) were performed in dichloromethane, and the maximum absorbance wavelength was used for the photoluminescence measurements (Figure 57 PL). All five target materials display two different absorbance bands, the first band in the region around 300 nm and the strongest absorbance band in the region of 350 nm. The first band at around 300 nm can be attributed to a π→π* transition located at the TPE core.[246] Interestingly, the band is observable for **TM_55**, **TM_56**, and **TM_59** (303 nm, 310 nm, and 309 nm), while the two target materials with a methoxy group in *ortho*-position only display the band as a shoulder to the main absorbance band (307 nm for **TM_57** and 284 nm for **TM_58**). Especially the strong blue shift of the shoulder for **TM_58** is of interest, as in this molecule, the methoxy groups are located at the *para*- and the *ortho*-position. The maximum absorbance band at around 350 nm can be attributed to an intramolecular charge transfer from the donor groups to the TPE core.[168, 246-247] Here, the two target materials with a methoxy group in the *ortho*-position (**TM_57** and **TM_58**) display the strongest blue shift with an absorbance band at 344 nm.

The second strongest blue shift is observable for **TM_56** at 351 nm, while the two target materials with a methoxy-group in the *para*-position display the strongest red shift at 358 nm for **TM_55** and 362 nm for **TM_59**.

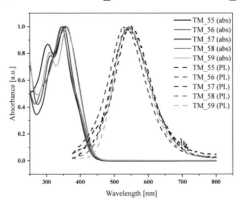

Figure 57: Normalized UV-vis spectra (in DCM, lines) and normalized photoluminescence spectra (in DCM, strokes) of the synthesized TPE derivatives.

Additionally, for all five target materials, photoluminescence spectra could be obtained. The strongest blue shift is observable for **TM_56** at 532 nm. The three target materials, **TM_57**, **TM_58**, and **TM_59**, display a photoluminescence maximum between 540 nm and 550 nm (542 nm, 546 nm, and 541 nm, respectively) and the *para*-methoxy-substituted **TM_55** shows the strongest red shift with 551 nm. The Stokes shifts between all target materials are substantial, with **TM_59** having the smallest Stokes shift of 179 nm, while the largest Stokes shift was observed for **TM_58** at 202 nm. These comparably large Stokes shifts for all molecules indicate large conformational changes in the excited state.[168] As the intersection point of the normalized UV-vis and photoluminescence spectra is observable, Equation 3 could be used to calculate the optical band gaps for all materials. The optical results are summarized in Table 22.

Table 22: Overview of optical properties of the TPE-based target materials.

Entry	Compound	λ_{max} [nm]	λ_{em} [nm]	Stokes shift [nm]	ΔE_g [eV]
1	TM_55	358	551	193	2.91
2	TM_56	351	532	190	2.99
3	TM_57	344	541	197	2.92
4	TM_58	344	546	202	2.87
5	TM_59	362	541	179	2.81

The calculated optical band gap of all five target materials ranges between 2.8 eV and 3.0 eV. The largest band gap is observed for the *meta*-substituted **TM_56,** while the *ortho-* and the *para*-substituted **TM_57** and **TM_55** obtain slightly smaller band gaps with 2.92 eV and 2.91 eV, respectively. The smallest band gaps are observed for the two target materials with more than one methoxy group, with **TM_58** (two methoxy groups: *ortho,* and *para*) having an optical band gap of 2.87 eV and **TM_59** (three methoxy groups: *meta, meta,* and *para*) obtaining a band gap of 2.81 eV. These changes could be due to their increased electron density and partly the mesomeric options of the *ortho-* and *para*-positions.

To determine the experimental HOMO and LUMO energy levels, cyclic voltammetry was conducted for the five target materials. The cyclic voltammo-grams of the TPE-based target materials are visualized in Figure 58 for the oxi-dative process and Figure 59 for the reductive process. The conditions can be found in chapter 3.2.2.

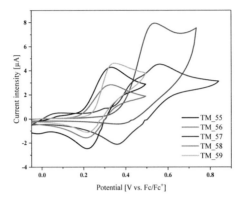

Figure 58: Oxidative cyclic voltammograms of the TPE-based target materials.

All five target materials show first oxidations between 0.2 v to 0.6 V (Figure 58). For **TM_55, TM_57,** and **TM_58,** the first oxidation curves are reversible, while the first oxidation process of **TM_56** is irreversible, and for **TM_59,** of quasi-reversible nature. The oxidation potentials of **TM_56** and **TM_57** are shifted to higher potentials compared to the other three target materials (~ 0.5 V *versus* ~ 0.3 V). The measured oxidation potentials *versus* the reference system ferrocene/ferrocenium are provided in Table 23. The corresponding reductive cyclic voltammograms are visualized in Figure 59.

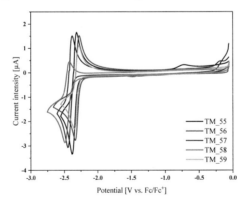

Figure 59: Reductive cyclic voltammograms of the TPE-based target materials.

All five compounds display a clear first reduction process in the range between -2.2 V and -2.7 V. The curves of the first reduction process are all reversible with the reduction process of **TM_58** having the highest negative potential while the lowest energy is required for **TM_56** and **TM_59**. The results of the electrochemical characterization are provided in Table 23.

Table 23: Electrochemical properties of the TPE-based target materials.

Entry	Compound	E_{ox} [V vs. Fc]	E_{red} [V vs. Fc]	E_{HOMO} [eV]	E_{LUMO} [eV]	$E_{HOMO-LUMO}$ [eV]	ΔE_g [eV]
1	TM_55	0.27	-2.35	-5.07	-2.45	2.62	2.91
2	TM_56	0.47	-2.30	-5.27	-2.50	2.75	2.99
3	TM_57	0.40	-2.42	-5.20	-2.39	2.81	2.92
4	TM_58	0.27	-2.46	-5.07	-2.35	2.72	2.87
5	TM_59	0.28	-2.31	-5.08	-2.49	2.59	2.81

The HOMO and LUMO energy levels were calculated according to Equation 4, and the results are provided in Table 23. While the HOMO energy levels of the three target materials with a methoxy-group in *para*-position are similar (-5.07 eV, -5.07 eV, and -5.08 eV for **TM_55**, **TM_56,** and **TM_57,** respectively), the HOMO energy levels of the two target materials containing the methoxy group in *meta*- (**TM_56**) and *ortho*-position (**TM_57**) are deeper by 0.13 eV and 0.20 eV respectively (Table 23, entries 2 and 3). In contrast to the HOMO energy levels, the LUMO energy levels follow a different trend. Here, the *ortho*-substitution of the methoxy groups seems to destabilize the LUMO compared to the other values. The LUMO energy levels of **TM_57** and **TM_58** are located at -2.39 eV and -2.35 eV, respectively, while the LUMO energy levels of the other three target materials are located at -2.45 cV for **TM_55**, -2.50 eV for **TM_56** and -2.49 eV for **TM_59**. These differences in HOMO and LUMO energy levels also lead to different band gaps with a similar trend when compared to the optical band gap. The smallest HOMO-LUMO band gaps are observed for the *para*-substituted target materials **TM_55**, **TM_58,** and **TM_59** (2.62 eV, 2.72 eV, and 2.59 eV, respectively). While **TM_56** and **TM_57** have the two largest band gaps (2.75 eV and 2.81 eV), the order is switched compared to the optical band gaps. However, as both calculations are approximations, the overall difference is not decisive. While theoretical calculations are ongoing, no results have been obtained yet.

Finally, the thermal decomposition of the five target materials was investigated *via* TGA (chapter 8.1, Figure 65). The different target materials begin decomposing between 280 °C and 350 °C, allowing the further application.

To summarize the results of this chapter, the absorbance maxima are blue-shifted enough not to have any competitive effects for the perovskite active materials. Also, the HOMO and LUMO energy levels of all five target materials are promising for further application in perovskite solar cells. The band gap of all target materials is between 2.6 eV and 3.0 eV (combined from optical and electrochemical characterization) and should therefore be large enough to prohibit any negative side effects like electron extraction, recombination, or electron transport. The target materials are under investigation regarding their applicability, but no results have been obtained yet.

3.6 MANUFACTURED DEVICES

3.6.1 Pseudo-*para*-substituted [2.2]Paracyclophanes

From the combined results of solubility and ionization potentials (see chapter 3.2.3), three target materials were chosen for further analysis and application into solar cells. The target materials **TM_31** (Di-*p*-OMeTPAThio-PCP), **TM_35** (Di-*p*-*tert*BuTPAThio-PCP), and **TM_37** (DiOMeCbzThio-PCP) all possess adequate solubilities and fitting HOMO and LUMO energy levels. The architecture of the used PSC is depicted in Figure 60a. The tin oxide was chosen as ETL, as the perovskite layer, the standard MAPbI$_3$ was introduced, and in addition to the HTL, a layer of MoO$_x$ was added for better hole extraction and transport between the HTM and the electrode.

The three target materials were all added *via* spin-coating, and the MoO$_x$ and silver electrodes were added *via* chemical vapor deposition (CVD). The I-V diagrams (see chapter 1.1) for the three materials are depicted in Figure 60b for **TM_31**, Figure 60c for **TM_35,** and Figure 60d for **TM_37**. All three target materials were added as pristine materials without any doping and the different curves in each diagram represent different irradiation, ranging from 10 W m^{-2} to the standard 1000 W m^{-2}. The different irradiations are chosen to further

116

characterize the efficiency of the hole transport materials, as the hole extraction capability of the HTM can be estimated from the form of the curve.

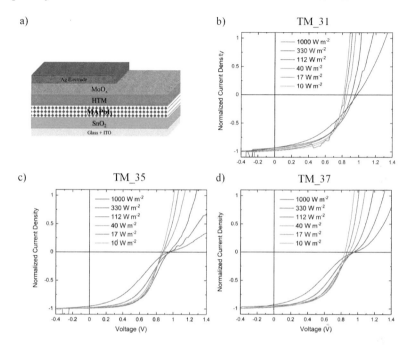

*Figure 60: a) Architecture of the perovskite solar cell. b) I-V diagram of **TM_31** at varying irradiance. C) I-V diagram of **TM_35** with varying irradiance. D) I-V diagram of **TM_37** with varying irradiance.*

The I-V diagrams of all three target materials contain a direct dependence between the curve and the chosen irradiation. This effect is strongest in **TM_35** and weakest in **TM_31**. The lower fill factor with stronger irradiance could be due to some extraction barriers between the perovskite and the hole transport layer. Therefore the I-V curve depicts better fill factors with lower irradiation, as the number of the obtained hole is lower, and a higher percentage can be extracted. The extraction barrier could be decreased by adding dopants such as LiTFSI and TBP to the hole transport layer, but while these experiments are running, no results can be provided yet. However, the PSC with the pristine target material **TM_31** as HTM reached a peak power conversion efficiency

117

under standard conditions of 8.4% and, therefore, is similar to pristine spiro-OMeTAD.[147]

In conclusion, three target materials were chosen as HTL in perovskite solar cells after further investigation, and the PSCs were tested under varying lighting conditions. This provided insights that there could be a hole extraction barrier between the perovskite and the HTL. This problem could be addressed by adding dopants to the HTL. Nonetheless, a PCE of 8.4% could be achieved with **TM_31** as the hole transport layer.

3.6.2 Bis(pseudo-*meta*)-*para*-substituted [2.2]Paracyclophanes

Only preliminary results for applying **TM_39** (tetra-*p*-OMeTPAThio-PCP) as a hole transport layer in a perovskite solar cell have been obtained. The test cell has been prepared in the same way and architecture as for the di-substituted target materials. Again, a layer of MoO_x was added between the organic HTM and the silver electrode for better hole transport to the electrode (Figure 61a). This additional layer allows the hole extraction barriers to be localized at the perovskite-HTM interface, not the HTM-electrode interface. Similar to the di-substituted target materials, different irradiation strengths were chosen to determine the effectivity of the hole extraction. The device architecture and the obtained I-V diagram are depicted in Figure 61b.

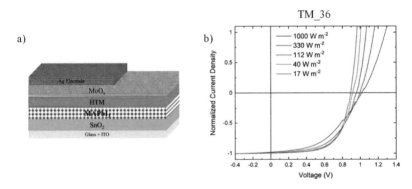

*Figure 61: a) Architecture of the perovskite solar cell. b) I-V diagram of **TM_39** at varying irradiance.*

The I-V diagram of **TM_39** displays a dependence between the curve and the irradiation, but the effect of the irradiation is less when compared to the di-substituted target materials (Figure 60b, c, and d). The decreased influence of the irradiation on the curve could imply that the tetra-substitution allows for a better interface contact with the increased number of arms and, therefore, decrease the hole extraction barrier. Additionally, the number of arms increases the inter-molecular hole transport inside the hole transport layer.[164] With an irradiance of 1000 W m^{-2}, the pristine target material **TM_39** achieved a peak power conversion efficiency of 11.0% and outperformed pristine spiro-OMeTAD in similar devices.[147] The influence of the dopants FK209, LiTFSI, and TBP is currently under investigation and could increase the hole extraction, hole mobility, and overall PCE.

In conclusion, three of the four investigated target materials displayed good solubilities and fitting ionization potentials. Target material **TM_39** was intro-duced into a perovskite solar cell as hole transport material and the following analysis provided insights that the hole extraction barrier between the perovskite and the hole transport layer is still there. However, increasing the arms from two to four decreased the hole extraction barrier while increasing the hole mobility slightly. **TM_40** and **TM_42** are currently under investigation as pristine materials, and for all three target materials, doping with FK209, LiTFSI, and TBP is studied. Without doping, **TM_39** could reach a peak PCE of 11.0%.

4 CONCLUSION AND OUTLOOK

In this thesis, 35 target materials were synthesized, of which 34 are unknown in the literature. The target materials' structure, purity, and optical and electrochemical properties were thoroughly characterized. Some of the most promising target materials were further analyzed, and for four target materials, the implementation as hole transport materials in perovskite solar cells could be presented.

Promising donor groups were identified in the first step, and ten donor groups were synthesized. While obtaining similar yields compared to the literature, the catalyst for the ULLMANN condensation could be switched from the air-instable copper(I) chloride to the more stable copper(I) iodide.

Then, as a first core building block, the pseudo-*para*-substituted [2.2]paracyclophane was chosen, and two different synthetical approaches were undertaken to synthesize the desired donor-π-bridge-substituted target materials. The idea behind the first inside-out approach was to build the target material starting from the core, attaching the π-bridge, and then attaching the donor moieties in a second step. This approach worked by implementing two consecutive CH activation reactions on the 2- and 5-position of the thiophene. However, the workup was strenuous, and the reaction time was long, so alternatives were developed. In the outside-in approach, the synthesized donor groups were first coupled with the borylated π-moieties (thiophene, EDOT, selenophene) *via* SUZUKI-MIYAURA cross-coupling conditions. In the second step, the coupling of the donor-π-moiety with the [2.2]paracylcophane was achieved *via* CH activation. By coupling the donor groups with the π-bridges first, nine new molecules were successfully synthesized, and for the already synthesized compounds, the yield could be increased while decreasing the reaction time. With the change from the inside-out approach to the outside-in approach, the workup was simplified, and the reaction time of the first step could be decreased from 18 hours to 2 hours and for the second step from 14 days to 18 hours.

Eight different target materials could be successfully synthesized, and after experimental characterization and theoretical calculations, three materials (**TM_31**, **TM_35,** and **TM_37**) could be implemented as hole transport materials into perovskite solar cells. While all three materials displayed some hole extraction barriers, a PCE of 8.4% could be achieved for the undoped **TM_31**. As all three target materials displayed similar hole extraction barriers, using dopants could drastically increase hole extraction and conductivity. Investigations with different possible dopants (LiTFSI, F4-TCNQ, TBP) are ongoing, and soon, more insights should be available. While the only target material containing EDOT as a π-bridge (**TM_33**) displayed some promising results in the electrochemical analysis, the solubility prohibited further investigation of the properties. Here, further modification of the donor groups (for example, using **28** as donor-π-bridge-moiety) could increase solubility and allow further applications.

With the second core building block chosen as the bis(pseudo-*meta*)-*para*-substituted [2.2]paracyclophane, the influence of increased substitution from two to four side arms should be investigated. Here, the most promising donor-π-bridge building blocks determined in the previous chapter were chosen. While the developed CH activation protocol did not provide promising results due to steric hindrance, a literature protocol applying a NEGISHI cross-coupling could be modified to fit the tetra-substitution on the [2.2]paracyclophane. The NEGISHI reagent could be synthesized *in situ,* and five different target materials could be obtained. After establishing the structure and purity, the electrochemical analysis showed that by increasing the conjugated π-system, the influence of the donors on the HOMO and LUMO energy levels was decreased, and a smaller deviation between the five target materials was achieved. The solubility of the five target materials increased compared to the di-substituted [2.2]paracyclophanes, especially the tetra-substituted **TM_40** dissolved better than its corresponding di-substituted **TM_32**. **TM_39** could be successfully implemented as hole transport material, and while displaying some hole extraction barriers, the peak power conversion efficiency could be increased to 11.0%. The investigation of undoped and doped charge carrier mobility is ongoing, and it is planned to introduce doped versions of all five synthesized target materials into perovskite solar cells for further investigation.

The number of side arms stayed the same in the third core building block, while the porphyrin macrocycle allowed further modification at the coordination site. For this investigation, the methoxy-containing starting materials **12, 14, 25, 29,** and **30** were chosen, and by applying the beforementioned NEGISHI cross-coupling conditions, five different zinc-containing target materials could be synthesized. After the solubility of **TM_48_Zn** was underwhelming, the other four compounds were demetalated following typical literature conditions, and the metalation with nickel acetate and cobalt acetate led to the synthesis of 17 different literature-unknown porphyrin macrocycles. After the structure and purity of each compound could be confirmed, the electrochemical analysis provided good insights regarding the influence of donor groups and metals. While the metals most strongly influenced the LUMO energy level, the variation between triarylamine and carbazole as donor groups led to differences of up to 0.3 eV between the related materials. This effect could further be confirmed by PESA analysis. Basic calculations of the structure and the HOMO and LUMO energy levels are ongoing. 16 of the 17 porphyrins displayed good solubility and thin-film morphology; further investigation is ongoing. It is planned to incorporate all 16 porphyrins into perovskite solar cells as hole transport materials. The effect of doping is still to be investigated, and due to the broad absorbance resulting from the Soret and Q bands, **TM_44_Zn** and **TM_45_Zn** are under investigation for possible use in organic solar cells.

As the last building block, tetraphenylethene was chosen. The basic structure of this core should allow easier investigation of the influence of the methoxy-substitution pattern of the donor groups on the overall properties. For this, synthesized donors **1**, **2**, **3**, **4**, and **5** were borylated following typical MIYAURA borylation conditions, and the core TPE was brominated (**49**). Following typical SUZUKI-MIYAURA cross-coupling conditions, the desired five target materials could be synthesized. After confirming the structure and purity of the compounds, the electrochemical analysis provided some insights into the effects of the methoxy groups. With a substitution in *para*-position, the HOMO energy levels were shallower while the band gap was smaller. However, the substitution in *ortho*-position destabilized the LUMO and increased the band gap, especially for **TM_57**. With fitting energy levels and good solubilities, the PESA measurements of the TPE-based materials are currently ongoing. It is planned to introduce the five materials into perovskite solar cells as a hole transport layer and further investigate the methoxy-substitution pattern's influence on their hole extraction and transport abilities.

To summarize the results obtained from the different synthesized materials, different trends could be observed regarding the HOMO and LUMO energy levels. When comparing the two employed donor groups, the triphenylamine and the carbazole, the carbazole-containing compounds contained deeper HOMO energy levels than their triphenylamine-containing analogs. A similar effect could be observed between the substituents on the donor groups. The methoxy groups provided the shallowest HOMO energy levels, while the *tert*-butyl-groups decreased the HOMO energy level minimally, and with no substituents, the deepest HOMO energy levels were obtained. Regarding the π-bridges, the HOMO energy level could be increased using EDOT or selenophene instead of thiophene. All in all, the different substituents on the various core building blocks allowed meticulous fine-tuning of the energy levels, and a range of the HOMO energy levels from -4.98 eV to -5.57 eV could be covered.

Generally, further investigation into the electronic properties is needed, especially theoretical ones. By simulating UV-vis spectra of the compounds, further insights into the excitation could be achieved, and the electron density of the HOMO and LUMO could provide more understanding into the exact effect of the different substituents, donor groups, π-bridges, and core molecules.

The synthesis, especially the π-bridges, could be modified by adding carbazoles and modifying the N-bound alkyl chains or by implementing dibenzothiophenes, thienothiophenes, or dithienothiophenes. Regarding porphyrins, literature has provided numerous examples of the influence of the different metals on charge conductivity, especially palladium, copper, and manganese offer interesting properties that could be further investigated. Instead of methoxy groups, the alkyl chain could be extended to increase the solubility further. For the tetra-phenylethenes, it would be interesting to see the difference in properties when fusing the different phenyl rings and increasing the rigidity of the core.

5 EXPERIMENTAL SECTION

5.1 GENERAL REMARKS

The synthetic procedures with all reaction details, as well as the analyses of the target materials, can be accessed additionally *via* the CHEMOTION repository:

https://www.chemotion-repository.net/home/publications

5.1.1 Materials and Methods

The starting materials, solvents, and reagents were purchased from ABCR, ACROS, ALFA AESAR, APOLLO SCIENTIFIC, CARBOLUTION, CHEMPUR, FLUOROCHEM, HONEYWELL, MERCK, SIGMA ALDRICH, TCI or THERMO FISHER SCIENTIFIC and used without further purification unless stated otherwise.

Solvents of technical quality were purified by distillation or with the solvent purification system MB SPS5 (dichloromethane, tetrahydrofuran, toluene) from MBRAUN. Solvents of *p.a.* quality were purchased from ACROS, FISHER SCIENTIFIC, SIGMA ALDRICH, ROTH, or RIEDEL-DEHAËN and were used without further purification. Other solvents were obtained from commercial suppliers: anhydrous N,N-dimethylformamide (SIGMA ALDRICH, <0.005% water), and anhydrous dimethyl sulfoxide (SIGMA ALDRICH, <0.005% water).

Oxygen-free solvents were obtained by degassing with argon for 20 minutes.

Liquids were added with a stainless-steel cannula, and solids were added in a powdered shape.

Air- and moisture-sensitive reactions were carried out under an argon atmosphere in oven-dried glassware using standard SCHLENK techniques.

Flat-bottomed crimp neck vials from CHROMAGLOBE with aluminum crimp caps were used for certain reactions.

Reactions at low temperatures were cooled using Dewars produced by ISOTHERM with water/ice or acetone/anhydrous ice mixtures.

Solvents were evaporated under reduced pressure at 45 °C using a rotary evaporator.

For solvent mixtures, each solvent was measured volumetrically.

Flash column chromatography was performed using MERCK silica 60 (0.040 x 0.063 mm, 230-400 mesh ASTM) and quartz sand (glowed and purified with hydrochloric acid). Alternatively, SIGMA ALDRICH aluminum oxide 150 basic (particle size: 70% between 0.063-0.200 mm) Brockmann activity III (6% H_2O (w/w)) was used.

All reactions were monitored by thin-layer chromatography (TLC) using silica-coated aluminum plates (MERCK, silica 60, F254). UV active compounds were detected with a UV lamp at 254 nm and 365 nm excitation. Basic potassium permanganate or Seebach solution was used as a TLC stain when required.

5.1.2 Devices

Nuclear Magnetic Resonance Spectroscopy (NMR)

NMR spectra were recorded on a BRUKER AVANCE 400 NMR instrument at 400 MHz for ^{1}H NMR, 100 MHz for ^{13}C NMR, and 128 MHz for ^{11}B NMR or a BRUKER AVANCE 500 NMR instrument at 500 MHz for ^{1}H NMR and 125 MHz for ^{13}C NMR. The NMR spectra were recorded at 22 °C in deuterated solvents acquired from EURISOTOP or DEUTERO. The chemical shift δ is displayed as parts per million [ppm] and the references used were the ^{1}H and ^{13}C peaks of the solvents: chloroform (^{1}H: δ = 7.27 ppm; ^{13}C: δ = 77.0 ppm), dichloromethane (^{1}H: δ = 5.32 ppm; ^{13}C: δ = 54.0 ppm) and tetrahydrofuran (^{1}H: δ = 3.58 ppm; ^{13}C: δ = 67.6 ppm). The recorded spectra were evaluated by 1st order. For the characterization of centrosymmetric signals, the median point was chosen, for multiplets in the signal range. The following abbreviations were used to describe the proton splitting pattern: s = singlet, d = doublet, t = triplet, m = multiplet, dd = doublet of a doublet, ddd = doublet of doublet of a doublet, dt = doublet of triplet, td = triplet of a doublet. Absolute values of the coupling constants J are given in Hertz [Hz]in absolute value and decreasing order. The assignment of the signals *via* ^{1}H NMR spectra was based on the multiplicity and the chemical shift. The assignment of the signals *via* ^{13}C NMR spectra was based on the chemical shift and the multiplicity obtained by phase-edited HSQC (*heteronuclear single quantum coherence*). Common solvent and solvent impurity signals as follows were not explicitly listed: chloroform: ^{1}H NMR 1.55

(H_2O), 1.25 (H grease), 0.84-0.87 (H grease), 0.07 (silicon grease) ppm; ^{13}C NMR 29.7 (H grease), 1.2 (silicon grease) ppm; dichloromethane: ^{1}H NMR 1.52 (H_2O), 1.29 (H grease), 0.84-0.90 (H grease), 0.09 (silicon grease) ppm; ^{13}C NMR 30.1 (H grease), 1.2 (silicon grease) ppm; tetrahydrofuran: ^{1}H NMR 10.84 (THF-d_8 impurity), 2.49 (H_2O), 1.73 (THF-d_8), 1.29 (H grease), 0.84-0.91 (H grease), 0.11 (silicon grease) ppm; ^{13}C NMR 29.9 (H grease), 1.2 (silicon grease) ppm.

Infrared Spectroscopy (IR)

The infrared spectra were recorded with a BRUKER ALPHA P instrument. All samples were measured by attenuated total reflection (ATR). The positions of the absorption bands are given in wavenumbers $\tilde{\nu}$ in cm^{-1} and were measured in the range from 3600 cm^{-1} to 500 cm^{-1}.

Mass Spectrometry (MS)

Electron ionization (EI) and fast atom bombardment (FAB) experiments were conducted using a FINNIGAN MAT 90 (70 eV) instrument, with 3-nitrobenzyl alcohol (3-NBA) as matrix and reference for high resolution. For the interpretation of the spectra, molecular peaks [M]$^+$, peaks of protonated molecules [M+H]$^+$, and characteristic fragment peaks are indicated with their mass-to-charge ratio (m/z) and their intensity in percent, relative to the base peak (100%), is given. In the case of high-resolution measurements, the maximum tolerated error is ±5 ppm.

Electron spray ionization (ESI) experiments were recorded on a THERMO FISHER SCIENTIFIC Q-EXACTIVE (ORBITRAP) mass spectrometer equipped with a HESI II probe to record high resolution or a THERMO FISHER SCIENTIFIC LTQ ORBITRAP XL. The tolerated error is ±5 ppm of the molecular mass. The spectra were interpreted by molecular peaks [M]$^+$, peaks of protonated molecules [M+H]$^+$, and characteristic fragment peaks and indicated with their mass-to-charge ratio (m/z).

Absorption Spectroscopy

UV-vis spectra were recorded on an ANALYTIK JENA SPECORD 50 Plus. Before the measurement, the samples were dissolved in dichloromethane or tetrahydro-furan and filled in glass cuvettes with a layer thickness of 0.1 cm.

Emission Spectroscopy

Emission spectra were recorded on a HORIBA SCIENTIFIC FLUOROMAX-4 spectro-fluorometer equipped with a CZERNY-TURNER-type monochromator and an R928P PMT detector.

Cyclic Voltammetry

Cyclic voltammetry experiments were performed with a GAMRY INTERFACE 1010B in a three electrodes electrochemical cell. The electrochemical cell was equipped with a glassy carbon (GC) working electrode, $Ag/AgNO_3$ reference electrode, and a platinum counter electrode.

The experiments were performed in N_2-saturated N,N-dimethylformamide containing 0.1 M [nBu$_4$N][PF$_6$] as the electrolyte at a scan rate of 100 mV s^{-1}. The concentration of the investigated compounds was 3 mM. According to the IUPAC recommendation, ferrocene (Fc) was added as an internal standard after each experiment.[248]

Melting Point

Melting points were measured on an OPTIMELT MPA100 device from STANFORD RESEARCH SYSTEM.

Thermogravimetric analysis (TGA)

Thermogravimetric analysis was performed on a TGA 5500 from TA INSTRUMENTS. The sample (~ 4 mg) was heated from 25 °C to 700 °C with a temperature increase of 25 °C min^{-1} on a platinum pan.

Theoretical calculations

They are performed by DAVID ELSING, working group WENZEL (KIT).

First, several conformers for each molecule were generated with force-field-based methods (CREST[249] with GFN-FF[250] for the disubstituted paracyclophanes and tetraphenylethenes, RDKit[251] with optimization by the MMFF94 force field for the tetrasubstituted paracyclophanes and OpenBabel[252] for the porphyrins). After filtering them by symmetry-corrected RMSD, they were optimized using DFTB+[253] with DFT-D3(BJ) dispersion correction[254-255], sorted by energy, and filtered again by symmetry-corrected RMSD before being optimized with DFT using the def2-SVP basis set and B3LYP functional with DFT-D3(BJ) dispersion correction using TURBOMOLE[256].

The electronic properties were then calculated on the optimized conformers with the def2-TZVP basis set and the B3LYP functional.

For the [2.2]paracyclophanes, only the pseudo-*para* conformers were used in the initial step, as was determined experimentally.

5.2 SYNTHESIS AND CHARACTERIZATION OF THE SYNTHESIZED COMPOUNDS

5.2.1 Synthesis of Donor Moieties

4-Bromo-*N,N*-bis(4-methoxyphenyl)aniline (1)

1-Iodo-4-methoxybenzene (8.16 g, 34.9 mmol, 3.00 equiv), 4-bromoaniline (2.00 g, 11.6 mmol, 1.00 equiv), CuI (57.6 mg, 581 μmol, 0.0500 equiv), 1,10-phenanthroline (105 mg, 581 μmol, 0.0500 equiv), KOH (5.22 g, 93.0 mmol, 8.00 equiv) and 60 mL of toluene were added to a flask with DEAN-STARK trap. The resulting solution was stirred at 135 °C for 48 h. The solvent was removed under reduced pressure, diluted with dichloromethane (300 mL), and washed with water (3x) and brine (3x). After drying over anhydrous Na$_2$SO$_4$, the solvent was removed under reduced pressure, and the crude product was purified *via* column chromatography on silica gel (eluent cyclohexane/dichloromethane 4:1) to yield 4-bromo-*N*,*N*-bis(4-methoxy-phenyl)aniline (4.14 g, 10.8 mmol, 93% yield) as an orange solid.

M.p.: 101 °C.

R_f = 0.6 (cyclohexane/ethyl acetate 4:1).

^{1}H NMR (400 MHz, Chloroform-d [7.27 ppm], ppm) δ = 7.25–7.22 (m, 2H), 7.06–7.02 (m, 4H), 6.85–6.79 (m, 6H), 3.80 (s, 6H).

^{13}C NMR (100 MHz, Chloroform-d [77.0 ppm], ppm) δ = 156.0 (2C), 147.9 (1C), 140.5 (2C), 131.7 (2C), 126.5 (4C), 122.0 (2C), 114.8 (4C), 112.3 (1C), 55.5 (2C).

IR (ATR, ṽ) = 3040, 2997, 2949, 2929, 2904, 2832, 1877, 1874, 1605, 1582, 1502, 1482, 1462, 1439, 1316, 1281, 1262, 1235, 1177, 1167, 1103, 1072, 1033, 1003, 953, 912, 820, 781, 730, 714, 694, 639, 602, 574, 534, 509, 490, 470, 442, 415 cm^{-1}.

FAB-MS (*m/z*, 3-NBA): 386/385/384/383 (25/97/27/100) [M]$^+$, HRMS–FAB (*m/z*): [M]$^+$ calcd for $C_{20}H_{18}O_2N_1{}^{79}Br_1$, 383.0515, found 383.0515.

4-Bromo-*N*,*N*-bis(3-methoxyphenyl)aniline (2)

1-Iodo-3-methoxy-benzene (4.07 g, 2.07 mL, 17.4 mmol, 2.20 equiv), 4-bromoaniline (1.36 g, 7.91 mmol, 1.00 equiv), CuI (113 mg, 593 µmol, 0.0750 equiv), 1,10-phenanthroline (107 mg, 593 µmol, 0.0750 equiv), KOH (3.55 g, 63.2 mmol, 8.00 equiv) and 40 mL of toluene were added to a flask with DEAN-STARK trap. The resulting solution was stirred at 135 °C for 4 days. The solvent was removed under reduced pressure, dissolved in 100 mL dichloromethane, and washed with water (3x) and brine (3x). After drying over anhydrous Na_2SO_4, the solvent was removed under reduced pressure, and the crude product was purified *via* column chromatography on silica gel (eluent cyclo-hexane/dichloromethane 2:1) to yield 4-bromo-*N*,*N*-bis(3-methoxy-phenyl)aniline (2.88 g, 7.49 mmol, 95% yield) as a purple solid.

M.p.: 70 °C.

R_f = 0.55 (cyclohexane/dichloromethane 1:1).

^{1}H NMR (400 MHz, Chloroform-d [7.27 ppm], ppm) δ = 7.34 (d, J = 8.9 Hz, 2H), 7.17 (t, J = 8.1 Hz, 2H), 6.97 (d, J = 8.8 Hz, 2H), 6.68–6.59 (m, 6H), 3.73 (s, 6H).

^{13}C NMR (100 MHz, Chloroform-d [77.0 ppm], ppm) δ = 160.5 (2C), 148.4 (2C), 146.7 (1C), 132.1 (2C), 129.9 (2C), 125.6 (2C), 116.9 (2C), 115.1 (1C), 110.2 (2C), 108.7 (2C), 55.2 (2C).

IR (ATR, ṽ) = 3067, 3002, 2955, 2830, 1577, 1483, 1465, 1448, 1431, 1401, 1317, 1283, 1272, 1258, 1244, 1208, 1169, 1149, 1118, 1105, 1086, 1071, 1037, 1000, 992, 960, 942, 882, 847, 839, 820, 785, 776, 761, 690, 636, 626, 596, 554, 506 cm^{-1}.

FAB-MS (m/z, 3-NBA): 385/383 (100/97) [M]$^+$, HRMS–FAB (m/z): [M]$^+$ calcd for $C_{20}H_{18}O_2N_1{}^{79}Br_1$, 383.0515, found 383.0513.

4-Bromo-N,N-bis(2-methoxyphenyl)aniline (3)

1-Iodo-2-methoxy-benzene (5.99 g, 3.33 mL, 25.6 mmol, 2.20 equiv), 4-bromoaniline (2.00 g, 11.6 mmol, 1.00 equiv), CuI (166 mg, 872 μmol, 0.0750 equiv), 1,10-phenanthroline (157 mg, 872 μmol, 0.0750 equiv), KOH (5.22 g, 93.0 mmol, 8.00 equiv) and 60 mL of toluene were added to a flask with DEAN-STARK trap. The resulting solution was stirred at 135 °C for 65 h. The solvent was removed under reduced pressure, and the crude product was purified *via* column chromatography on silica gel (eluent cyclohexane/ethyl acetate 20:1) to yield 4-bromo-N,N-bis(2-methoxyphenyl)aniline (2.36 g, 6.14 mmol, 53% yield) as a red solid.

M.p.: 114 °C.

R_f = 0.65 (cyclohexane/ethyl acetate 4:1).

^{1}H NMR (400 MHz, Chloroform-d [7.27 ppm], ppm) δ = 7.20–7.17 (m, 6H), 6.97–6.90 (m, 4H), 6.51 (d, J = 8.9 Hz, 2H), 3.70 (s, 6H).

^{13}C NMR (100 MHz, Chloroform-d [77.0 ppm], ppm) δ = 155.3 (2C), 147.4 (1C), 134.9 (2C), 131.1 (2C), 128.8 (2C), 126.4 (2C), 121.3 (2C), 118.4 (2C), 112.7 (2C), 110.8 (1C), 55.7 (2C).

IR (ATR, ṽ) = 3060, 3016, 2956, 2931, 2878, 2832, 1582, 1493, 1485, 1455, 1434, 1330, 1313, 1296, 1288, 1265, 1234, 1179, 1159, 1122, 1112, 1072, 1044, 1021, 1000, 972, 945, 915, 851, 813, 778, 748, 738, 710, 690, 632, 619, 564, 540 cm^{-1}.

FAB-MS (m/z, 3-NBA): 385/383 (95/100) [M]$^+$, 279/277 (47/44); HRMS–FAB (m/z): [M]$^+$ calcd for $C_{20}H_{18}O_2N_1{}^{79}Br_1$, 383.0515, found 383.0517.

4-Bromo-N,N-bis(2,4-dimethoxyphenyl)aniline (4)

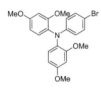 4-Bromoaniline (1.50 g, 8.72 mmol, 1.00 equiv), 1-iodo-2,4-dimethoxy-benzene (5.07 g, 19.2 mmol, 2.20 equiv), CuI (125 mg, 654 μmol, 0.0750 equiv), 1,10-phenanthroline (118 mg, 654 μmol, 0.0750 equiv), and KOH (3.91 g, 69.8 mmol, 8.00 equiv) were added to a flask with DEAN-STARK trap and 45 mL of toluene were added. The resulting solution was stirred at 135 °C for 72 h. The solvent was removed under reduced pressure, and the crude product was purified *via* column chromatography on silica gel (eluent cyclohexane/ethyl acetate 20:1 → 10:1) to yield 4-bromo-N,N-bis(2,4-dimethoxy-phenyl)aniline (2.17 g, 4.88 mmol, 56% yield) as a red solid.

M.p.: 138 °C.

R_f = 0.3 (cyclohexane/ethyl acetate 4:1).

^{1}H NMR (400 MHz, Chloroform-d [7.27 ppm], ppm) δ = 7.22 (d, J = 8.6 Hz, 2H), 7.14 (d, J = 8.9 Hz, 2H), 6.53 (d, J = 2.5 Hz, 2H), 6.46 (dd, J = 2.6 Hz, J = 8.6 Hz, 2H), 6.36 (d, J = 8.8 Hz, 2H), 3.81 (s, 6H), 3.74 (s, 6H).

^{13}C NMR (100 MHz, Chloroform-d [77.0 ppm], ppm) δ = 158.8 (2C), 156.8 (2C), 148.2 (1C), 131.0 (2C), 130.3 (2C), 127.7 (2C), 116.1 (2C), 109.1 (1C), 104.8 (2C), 100.0 (2C), 55.7 (2C), 55.4 (2C).

IR (ATR, ṽ) = 3009, 2951, 2929, 2905, 2883, 2833, 1602, 1581, 1506, 1485, 1462, 1434, 1412, 1340, 1322, 1300, 1275, 1242, 1205, 1184, 1166, 1159, 1126, 1109, 1071, 1031, 999, 993, 949, 936, 909, 898, 890, 844, 823, 810, 741, 731, 722, 691, 636, 599, 582, 567, 541 cm^{-1}.

FAB-MS (m/z, 3-NBA): 445/443 (99/100) [M]$^+$; HRMS–FAB (m/z): [M]$^+$ calcd for $C_{22}H_{22}O_4N_1{}^{79}Br_1$, 443.0727, found 443.0729.

4-Bromo-*N,N*-bis(3,4,5-trimethoxyphenyl)aniline (5)

4-Bromoaniline (1.30 g, 7.56 mmol, 1.00 equiv), 5-iodo-1,2,3-trimethoxy-benzene (4.89 g, 16.6 mmol, 2.20 equiv), CuI (108 mg, 567 μmol, 0.0750 equiv), 1,10-phenanthroline (102 mg, 567 μmol, 0.0750 equiv), KOH (3.39 g, 60.5 mmol, 8.00 equiv) and 40 mL toluene were added to a flask with DEAN-STARK trap. The resulting solution was stirred at 135 °C for 68 h. The solvent was removed under reduced pressure, and the crude product was purified *via* column chromatography on silica gel (eluent cyclohexane/ethyl acetate 10:1 + 1% triethylamine) to yield 4-bromo-*N,N*-bis(3,4,5-trimethoxyphenyl)aniline (3.05 g, 6.05 mmol, 80% yield) as an off-white solid.

M.p.: 125 °C.

R_f = 0.21 (cyclohexane/ethyl acetate 4:1).

^{1}H NMR (400 MHz, Chloroform-d [7.27 ppm], ppm) δ = 7.32 (d, *J* = 8.8 Hz, 2H), 6.94 (d, *J* = 8.8 Hz, 2H), 6.29 (s, 4H), 3.85 (s, 6H), 3.73 (s, 12H).

^{13}C NMR (100 MHz, Chloroform-d [77.0 ppm], ppm) δ = 153.7 (4C), 146.9 (1C), 143.1 (2C), 134.4 (2C), 131.9 (2C), 124.4 (2C), 114.3 (1C), 102.3 (4C), 61.0 (2C), 56.2 (4C).

IR (ATR, ṽ) = 2997, 2973, 2953, 2932, 2900, 2885, 2829, 1584, 1499, 1485, 1446, 1417, 1401, 1349, 1302, 1271, 1224, 1208, 1181, 1125, 1069, 1050, 1024, 999, 921, 833, 807, 788, 768, 751, 738, 720, 707, 687, 652, 619, 501, 459 cm^{-1}.

FAB-MS (*m/z*, 3-NBA): 507/506/505/504/503 (8/40/100/41/98) [M]$^+$, 491/490/489/488 (9/22/10/22); HRMS–FAB (*m/z*): [M]$^+$ calcd for $C_{24}H_{26}O_6N_1{}^{79}Br_1$, 503.0938, found 503.0936.

4-Bromo-*N,N*-bis(4-*tert*-butylphenyl)aniline (6)

4-Bromoaniline (1.50 g, 8.72 mmol, 1.00 equiv), 1-tert-butyl-4-iodobenzene (4.99 g, 19.2 mmol, 2.20 equiv), CuCl (39.9 mg, 403 µmol, 0.0462 equiv), 1,10-phenanthroline (78.6 mg, 436 µmol, 0.0500 equiv), KOH (3.91 g, 69.8 mmol, 8.00 equiv) and 45 mL of toluene were added to a flask with DEAN-STARK trap. The resulting solution was stirred at 135 °C for 65 h. The solvent was removed under reduced pressure, diluted with dichloromethane (300 mL), and washed with water (3x 100 mL) and brine (3x 100 mL). After drying over anhydrous Na_2SO_4, the solvent was removed under reduced pressure, and the crude product was purified *via* column chromatography on silica gel (eluent n pentane → cyclohexane) to yield 4-bromo-*N,N*-bis(4-*tert*-butylphenyl)aniline (3.10 g, 7.10 mmol, 81% yield) as a light brown solid.

M.p.: 166 °C.

R_f = 0.55 (*n*-pentane)

^{1}H NMR (400 MHz, Chloroform-d [7.27 ppm], ppm) δ = 7.32–7.27 (m, 6H), 7.03–7.01 (m, 4H), 6.97–6.93 (m, 2H), 1.34 (s, 18H).

^{13}C NMR (100 MHz, Chloroform-d [77.0 ppm], ppm) δ = 147.3 (1C), 146.0 (2C), 144.6 (2C), 131.9 (2C), 126.1 (4C), 124.3 (2C), 124.0 (4C), 113.7 (1C), 34.3 (2C), 31.4 (6C).

IR (ATR, ṽ) = 3040, 2959, 2900, 2864, 1602, 1579, 1507, 1483, 1460, 1391, 1361, 1322, 1295, 1283, 1268, 1201, 1191, 1177, 1109, 1069, 1014, 1006, 827, 816, 749, 735, 710, 693, 568, 554, 541 cm^{-1}.

FAB-MS (*m/z*, 3-NBA): 437/435 (99/100) [M]$^+$, 431 (23), 422/420 (28/29); HRMS–FAB (*m/z*): [M]$^+$ calcd for $C_{26}H_{30}N_1{}^{79}Br_1$, 435.1556, found 435.1554.

4-Bromo-*N*,*N*-bis(4-methylthiophenyl)aniline (7)

4-Bromoaniline (1.50 g, 8.72 mmol, 1.00 equiv), 1-iodo-4-thioanisole (4.80 g, 19.2 mmol, 2.20 equiv), CuI (125 mg, 654 µmol, 0.0750 equiv), 1,10-phenanthroline (118 mg, 654 µmol, 0.0750 equiv), KOH (3.91 g, 69.8 mmol, 8.00 equiv) and 40 mL of toluene were added to a flask with DEAN-STARK trap. The resulting solution was stirred at 135 °C for 5.7 days. The solvent was removed under reduced pressure, dissolved in dichloromethane (300 mL), and washed with water (3x 200 mL) and brine (3x 200 mL). After drying over anhydrous Na_2SO_4, the solvent was removed under reduced pressure, and the crude product was purified *via* column chromatography (eluent cyclo-hexane/ethyl acetate 200:1 + 1% triethylamine) to yield 4-bromo-*N*,*N*-bis(4-methylthiophenyl)aniline (3.08 g, 7.39 mmol, 85% yield) as a brown oil.

M.p.: 73 °C.

R_f = 0.65 (cyclohexane/ethyl acetate 50:1).

^{1}H NMR (400 MHz, Chloroform-d [7.27 ppm], ppm) δ = 7.32 (d, *J* = 8.8 Hz, 2H), 7.18 (d, *J* = 8.6 Hz, 4H), 7.00 (d, *J* = 8.6 Hz, 4H), 6.93 (d, *J* = 8.8 Hz, 2H), 2.48 (s, 6H).

^{13}C NMR (100 MHz, Chloroform-d [77.0 ppm], ppm) δ = 146.6 (1C), 144.8 (2C), 132.4 (2C), 132.2 (2C), 128.5 (4C), 124.8 (4C), 124.8 (2C), 114.8 (1C), 16.7 (2C).

IR (ATR, ṽ) = 3023, 2979, 2917, 2847, 2829, 2285, 1890, 1579, 1482, 1435, 1425, 1397, 1307, 1281, 1266, 1194, 1174, 1123, 1106, 1094, 1069, 1004, 965, 953, 915, 813, 754, 714, 683, 674, 635, 521, 510 cm^{-1}.

FAB-MS (*m/z*, 3-NBA): 418/417/416/415 (31/100/34/93) [M]$^+$, HRMS–FAB (*m/z*): [M]$^+$ calcd for $C_{20}H_{18}N_1{}^{79}Br_1S_2$, 415.0059, found 415.0059.

3,6-Dimethoxy-9H-carbazole (8)

3,6-Dibromo-9H-carbazole (4.00 g, 12.3 mmol, 1.00 equiv), CuI (9.87 g, 51.8 mmol, 4.21 equiv), NaOMe (13.3 g, 45.6 mL, 246 mmol, 5.40M, 20.0 equiv) and dry N,N-dimethylformamide (133 ml) were added to a flask under argon atmosphere and heated from 90 °C to 140 °C over the course of 2 h. The reaction mixture was kept at 140 °C for 2 hours, then cooled to 22 °C. After adding ethyl acetate (250 ml), the mixture was filtered over celite and washed with brine (3 x 150 ml). The solvent was removed under reduced pressure. The crude product was purified *via* column chromatography on silica gel (eluent cyclohexane/ethyl acetate 8:1 → 4:1) to yield 3,6-dimethoxy-9H-carbazole (2.56 g, 11.3 mmol, 92% yield) as a colorless solid.

M.p.: 102 °C.

R_f = 0.45 (cyclohexane/ethyl acetate 4:1).

^{1}H NMR (400 MHz, Chloroform-d [7.27 ppm], ppm) δ = 7.78 (s, 1H), 7.52 (d, J = 2.5 Hz, 2H), 7.30–7.27 (m, 2H), 7.07 (dd, J = 2.5 Hz, J = 8.8 Hz, 2H), 3.95 (s, 6H).

^{13}C NMR (100 MHz, Chloroform-d [77.0 ppm], ppm) δ = 153.6 (2C), 135.2 (2C), 123.7 (2C), 115.2 (2C), 111.5 (2C), 102.8 (2C), 56.0 (2C).

IR (ATR, ṽ) = 3357, 2992, 2961, 2941, 2836, 1857, 1611, 1575, 1494, 1482, 1468, 1451, 1435, 1333, 1302, 1281, 1265, 1230, 1205, 1179, 1149, 1122, 1111, 1021, 932, 909, 843, 807, 775, 737, 730, 642, 595, 558 cm^{-1}.

EI (*m/z*, 70 eV, 70 °C): 228/227 (12/100) [M]$^+$, 213/212 (14/95), 197 (10), 184 (24), 141/140 (30/14), 69 (15); HRMS–EI (*m/z*): [M]$^+$ calcd for $C_{14}H_{13}N_1O_2$, 227.0946; found, 227.0941.

9-(4-Bromophenyl)-3,6-dimethoxycarbazole (9)

3,6-Dimethoxy-9H-carbazole (8) (2.20 g, 9.68 mmol, 1.00 equiv), 1,4-dibromobenzene (9.13 g, 38.7 mmol, 4.00 equiv), CuI (880 mg, 4.62 mmol, 0.477 equiv) and K_2CO_3 (4.01 g, 29.0 mmol, 3.00 equiv) were added into a 250 mL flask. Dry *N,N*-dimethylacetamide (70 mL) was added to the flask. The solution was kept at 180 °C and stirred vigorously under an argon atmosphere for 44 h. After cooling to 22 °C, the solution was poured into water (100 mL) and extracted with dichloromethane (3x 100 mL). The combined organic phases were washed with water (3x 250 mL) and brine (3x 200 mL), dried over anhydrous Na_2SO_4, and the solvent was removed under reduced pressure. The obtained crude product was purified *via* column chromatography on silica gel (eluent cyclohexane/ethyl acetate 40:1 → 10:1) to yield 9-(4-bromophenyl)-3,6-dimethoxycarbazole (2.88 g, 7.53 mmol, 78% yield) as a colorless solid.

M.p.: 136 °C.

R_f = 0.25 (cyclohexane/ethyl acetate 40:1).

^{1}H NMR (400 MHz, Chloroform-d [7.27 ppm], ppm) δ = 7.72–7.69 (m, 2H), 7.56 (d, *J* = 2.5 Hz, 2H), 7.45–7.42 (m, 2H), 7.31 (d, *J* = 8.8 Hz, 2H), 7.05 (dd, *J* = 2.5 Hz, *J* = 8.8 Hz, 2H), 3.96 (s, 6H).

^{13}C NMR (100 MHz, Chloroform-d [77.0 ppm], ppm) δ = 154.2 (2C), 137.3 (1C), 135.9 (2C), 133.0 (2C), 128.2 (2C), 123.8 (2C), 120.2 (1C), 115.3 (2C), 110.5 (2C), 103.0 (2C), 56.1 (2C).

IR (ATR, ṽ) = 3070, 3002, 2990, 2949, 2931, 2895, 2829, 1604, 1581, 1570, 1487, 1469, 1455, 1441, 1428, 1400, 1373, 1327, 1305, 1289, 1261, 1249, 1203, 1177, 1153, 1106, 1068, 1028, 1009, 956, 945, 929, 909, 854, 836, 823, 802, 789, 754, 737, 721, 711, 657, 628, 601, 586 cm^{-1}.

FAB-MS (*m/z*, 3-NBA): 383/381 (100/99) [M]$^+$, 154 (27), 136 (31); HRMS–FAB (*m/z*): [M]$^+$ calcd for $C_{20}H_{16}O_2N_1{}^{79}Br_1$, 381.0359, found 381.0357.

9-(4-Bromophenyl)-3,6-di*tert*-butylcarbazole (10)

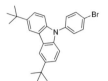 3,6-Di*tert*-butyl-9H-carbazole (3.00 g, 10.7 mmol, 1.00 equiv) and Cs₂CO₃ (7.00 g, 21.5 mmol, 2.00 equiv) were added to a 20 mL crimp cap vial and evacuated and refilled with argon three times. 1-bromo-4-fluorobenzene (3.76 g, 2.36 mL, 21.5 mmol, 2.00 equiv) and 12 mL *N,N*-di-methylformamide were added *via* syringe, and the reaction mixture was heated to 150 °C for 69 h. The mixture was cooled to 22 °C, added to 80 mL brine, and extracted with dichloromethane (3x 50 mL). The combined organic phases were dried over anhydrous Na₂SO₄, and the solvent was removed under reduced pressure. The crude product was purified *via* column chromatography on silica gel (eluent cyclohexane/ethyl acetate 50:1) to yield 9-(4-bromophenyl)-3,6-di*tert*-butylcarbazole (4.20 g, 9.67 mmol, 90% yield) as a colorless solid.

M.p.: 154 °C.

R_f = 0.85 (cyclohexane/ethyl acetate 4:1).

¹H NMR (400 MHz, Chloroform-d [7.27 ppm], ppm) δ = 8.15–8.15 (m, 2H), 7.73–7.70 (m, 2H), 7.49–7.44 (m, 4H), 7.33 (d, *J* = 8.6 Hz, 2H), 1.48 (s, 18H).

¹³C NMR (100 MHz, Chloroform-d [77.0 ppm], ppm) δ = 143.2 (2C), 138.9 (2C), 137.3 (1C), 133.0 (2C), 128.3 (2C), 123.7 (2C), 123.5 (2C), 120.2 (1C), 116.3 (2C), 109.0 (2C), 34.7 (2C), 32.0 (6C).

IR (ATR, ṽ) = 3061, 2962, 2949, 2902, 2864, 1623, 1585, 1489, 1470, 1363, 1326, 1296, 1259, 1235, 1193, 1170, 1102, 1067, 1035, 1010, 881, 840, 822, 810, 765, 741, 720, 694, 652, 632, 611, 595, 500 cm⁻¹.

FAB-MS (*m/z*, 3-NBA): 435/433 (100/97) [M]⁺, 420 (37), 418 (39), 380 (13), 378 (15), 154 (23), 136 (19); HRMS–FAB (*m/z*): [M]⁺ calcd for C₂₆H₂₈N₁⁷⁹Br₁, 433.1400, found 433.1398.

5.2.2 Pseudo-*para*-substituted [2.2]Paracyclophanes

4,16-Dithienyl-[2.2]paracyclophane (11)

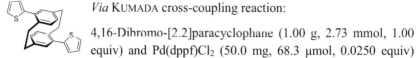

Via KUMADA cross-coupling reaction:

4,16-Dibromo-[2.2]paracyclophane (1.00 g, 2.73 mmol, 1.00 equiv) and Pd(dppf)Cl$_2$ (50.0 mg, 68.3 µmol, 0.0250 equiv) were suspended in 20 mL of dry tetrahydrofuran and degassed with argon for 15 minutes. Thienyl magnesium bromide (2.05 g, 10.9 mL, 10.9 mmol, 1.00M in THF, 4.00 equiv) was added *via* syringe, and the reaction mixture was heated to 60 °C for 2 h. The reaction progress was monitored *via* TLC. The product mixture was cooled to 22 °C, and saturated aqueous NH$_4$Cl was added (10 mL). The organic phase was diluted with 20 mL dichloromethane; the phases were separated and washed with 2M HCl. The solvent was removed under reduced pressure, and the crude product was washed with cyclohexane and cold dichloromethane to yield 4,16-dithienyl-[2.2]paracyclophane (193 mg, 517 µmol, 19% yield) as a colorless solid.

Via CH activation:

4,16-Dibromo-[2.2]paracyclophane (500 mg, 1.37 mmol, 1.00 equiv), Pd(OAc)$_2$ (15.3 mg, 68.3 µmol, 0.0500 equiv), PPh$_3$ (35.8 mg, 137 µmol, 0.100 equiv), PivOH (41.8 mg, 410 µmol, 0.300 equiv) and Cs$_2$CO$_3$ (667 mg, 2.05 mmol, 1.50 equiv) were suspended in 6 mL of dry *N,N*-dimethylformamide in an argon atmosphere. Thiophene (2.30 g, 2.17 mL, 27.3 mmol, 20.0 equiv) was added *via* syringe, and the reaction mixture was heated to 100 °C for 48 h. The reaction progress was monitored *via* TLC, and after 48 h, the product mixture was cooled to 22 °C and poured into 20 mL of water. The mixture was extracted with dichloromethane (3x 20 mL), and the combined organic phases were washed with water (3x 50 mL) and brine (2x 50 mL), dried over anhydrous Na$_2$SO$_4$ and the solvent was removed under reduced pressure. The obtained crude product was purified *via* column chromatography on silica gel (eluent cyclohexane/ethyl acetate 50:1 → 10:1). Afterwards, the product was recrystallized from dimethyl sulfoxide to yield 4,16-dithienyl-[2.2]paracyclophane (360 mg, 966 µmol, 71% yield) as a colorless solid.

M.p.: 292 °C.

$R_f = 0.2$ (cyclohexane/ethyl acetate 20:1).

^{1}H NMR (400 MHz, Chloroform-d [7.27 ppm], ppm) δ = 7.41 (dd, J = 1.3 Hz, J = 5.1 Hz, 2H), 7.18–7.14 (m, 4H), 6.76 (dd, J = 1.9 Hz, J = 7.7 Hz, 2H), 6.68 (d, J = 1.8 Hz, 2H), 6.61 (d, J = 7.8 Hz, 2H), 3.78–3.72 (m, 2H), 3.03–2.96 (m, 2H), 2.92–2.88 (m, 4H).

^{13}C NMR (100 MHz, Chloroform-d [77.0 ppm], ppm) δ = 144.2 (2C), 140.1 (2C), 137.2 (2C), 134.9 (2C), 134.9 (2C), 133.5 (2C), 129.6 (2C), 127.6 (2C), 126.0 (2C), 125.2 (2C), 34.4 (2C), 34.0 (2C).

IR (ATR, $\tilde{v}$) = 2962, 2934, 2891, 2850, 1587, 1523, 1476, 1452, 1434, 1400, 1349, 1255, 1210, 1193, 1154, 1077, 1048, 1010, 949, 897, 864, 851, 820, 747, 731, 684, 654, 622, 591, 541, 514 cm^{-1}.

EI-MS (m/z, 70 eV, 70 °C): 372 (65) [M]$^+$, 187 (100), 186 (64), 185 (76), 184 (27), 171 (45), 141 (20); HRMS–EI (m/z): [M]$^+$ calcd for $C_{24}H_{20}{}^{32}S_2$, 372.1000; found, 372.1000.

4-(2-Thienyl)-N,N-bis(4-methoxyphenyl)aniline (12)

Via CH activation:

4-Bromo-N,N-bis(4-methoxyphenyl)aniline (1) (1.00 g, 2.60 mmol, 1.00 equiv), PEPPSI™-iPr (88.7 mg, 130 µmol, 0.0500 equiv), P(Cy)$_3$ (104 mg, 260 µmol, 0.100 equiv), K$_2$CO$_3$ (539 mg, 3.90 mmol, 1.50 equiv) and PivOH (79.7 mg, 781 µmol, 0.300 equiv) were added to a 50 mL crimp cap vial and evacuated and backfilled with argon. Thiophene (2.19 g, 2.09 mL, 26.0 mmol, 10.0 equiv) and dry N,N-dimethylformamide (33.4 mL) were added *via* syringe, and the reaction mixture was heated to 100 °C under constant stirring. The reaction progress was monitored *via* TLC. After 45 h, the product mixture was cooled to 22 °C, poured into 100 mL water, and extracted with ethyl acetate (3x 100 mL). The combined organic phases were washed with water (3x 100 mL) and brine (3x 100 mL), dried over anhydrous Na$_2$SO$_4$, and the solvent was removed under reduced pressure. The obtained crude product was purified *via* column chromatography on silica gel (eluent cyclohexane/ethyl acetate 20:1) to yield 4-(2-thienyl)-N,N-bis(4-methoxyphenyl)aniline (761 mg, 1.96 mmol, 75% yield) as an orange solid.

Via SUZUKI-MIYAURA cross-coupling:

4-Bromo-*N,N*-bis(4-methoxyphenyl)aniline (1) (1.50 g, 3.90 mmol, 1.00 equiv), 2-pinacolborylthiophene (15) (984 mg, 4.68 mmol, 1.20 equiv), Pd(dppf)Cl$_2$ (143 mg, 195 μmol, 0.0500 equiv) and K$_2$CO$_3$ (3.24 g, 23.4 mmol, 6.00 equiv) were added to a 50 mL crimp cap vial and refilled with argon three times. 40 mL of degassed dimethyl sulfoxide was added, and the reaction mixture was heated to 90 °C. The reaction progress was monitored *via* TLC, and after 4.5 h, the product mixture was cooled to 22 °C. The mixture was poured into 100 mL water and extracted with ethyl acetate (3x 100 mL). The combined organic phases were washed with water (3x 200 mL) and brine (3x 200 mL), dried over an-hydrous Na$_2$SO$_4$, and the solvents were removed under reduced pressure. The crude product was purified *via* column chromatography on silica gel (eluent cyclohexane/ethyl acetate 20:1) to yield 4-(2-thienyl)-*N,N*-bis(4-methoxy-phenyl)aniline (1.28 g, 3.29 mmol, 84% yield) as an orange solid.

M.p.: 117 °C.

R_f = 0.6 (cyclohexane/ethyl acetate 4:1).

^{1}H NMR (400 MHz, Dichloromethane-d2 [5.32 ppm], ppm) δ = 7.43–7.39 (m, 2H), 7.21–7.19 (m, 2H), 7.08–7.03 (m, 5H), 6.90–6.87 (m, 2H), 6.87–6.83 (m, 4H), 3.79 (s, 6H).

^{13}C NMR (100 MHz, Dichloromethane-d2 [54.0 ppm], ppm) δ = 156.7 (2C), 148.9 (1C), 145.1 (1C), 141.1 (2C), 128.5 (1C), 127.3 (4C), 127.0 (2C), 126.9 (1C), 124.0 (1C), 122.2 (1C), 120.8 (2C), 115.2 (4C), 56.0 (2C).

IR (ATR, ṽ) = 3105, 3012, 2945, 2922, 2900, 2832, 1871, 1731, 1599, 1534, 1494, 1453, 1439, 1434, 1425, 1402, 1354, 1332, 1298, 1278, 1266, 1241, 1191, 1183, 1160, 1130, 1101, 1082, 1035, 1007, 955, 936, 909, 844, 826, 812, 782, 737, 715, 696, 646, 642, 632, 595, 569, 537, 527 cm^{-1}.

FAB-MS (*m/z*, 3-NBA): 389/388/387 (12/39/100) [M]$^+$; HRMS FAB (*m/z*): [M]$^+$ calcd for C$_{24}$H$_{21}$O$_2$N$_1$S$_1$, 387.1288, found 387.1290.

4-(2-Thienyl)-*N*,*N*-bis(phenyl)aniline (13)

Via CH activation:

4-Bromo-*N*,*N*-bisphenylaniline (500 mg, 1.54 mmol, 1.00 equiv), PEPPSI™-iPr (52.5 mg, 77.1 µmol, 0.0500 equiv), P(Cy)$_3$ (61.8 mg, 154 µmol, 0.100 equiv), K$_2$CO$_3$ (320 mg, 2.31 mmol, 1.50 equiv) and PivOH (47.3 mg, 463 µmol, 0.300 equiv) were added to a 20 mL crimp cap vial and evacuated and backfilled with argon. Thiophene (195 mg, 185 µL, 2.31 mmol, 1.50 equiv) and dry *N*,*N*-dimethylformamide (15.0 mL) were added *via* syringe, and the reaction mixture was heated to 100°C under constant stirring. The reaction progress was monitored *via* TLC. After 4 h, the product mixture was cooled to 22 °C, poured into 50 mL water, and extracted with dichloromethane (3x 50 mL). The combined organic phases were washed with water (3x 50 mL) and brine (3x 50 mL), dried over anhydrous Na$_2$SO$_4$, and the solvent was removed under reduced pressure. The crude product was purified *via* column chromatography on silica gel (eluent cyclohexane/ethyl acetate 4:1) to yield 4-(2-thienyl)-*N*,*N*-bis(phenyl)aniline (441 mg, 1.35 mmol, 87% yield) as a dark solid.

Via SUZUKI-MIYAURA cross-coupling:

4-Bromo-*N*,*N*-bisphenylaniline (1.00 g, 3.08 mmol, 1.00 equiv), 4,4,5,5-tetra-methyl-2-thiophen-2-yl-1,3,2-dioxaborolane (778 mg, 3.70 mmol, 1.20 equiv), Pd(dppf)Cl$_2$ (113 mg, 154 µmol, 0.0500 equiv) and K$_2$CO$_3$ (2.56 g, 18.5 mmol, 6.00 equiv) were added to a 50 mL crimp cap vial and refilled with argon three times. 35 mL of degassed dimethyl sulfoxide was added, and the reaction mixture was heated to 90 °C. The reaction progress was monitored *via* TLC, and after 1.5 h, the product mixture was cooled to 22 °C. The mixture was poured into 100 mL water and extracted with ethyl acetate (3x 100 mL). The combined organic phases were washed with water (3x 200 mL) and brine (3x 200 mL), dried over anhydrous Na$_2$SO$_4$, and the solvent was removed under reduced pressure. The crude product was purified *via* column chromatography on silica gel (eluent cyclohexane/ethyl acetate 5:1) to yield 4-(2-thienyl)-*N*,*N*-bis(phenyl)aniline (1.00 g, 3.07 mmol, 100% yield) as a dark solid.

M.p.: 124 °C.

R_f = 0.85 (cyclohexane/ethyl acetate 4:1).

^{1}H NMR (400 MHz, Dichloromethane-d2 [5.32 ppm], ppm) δ = 7.51–7.47 (m, 2H), 7.30–7.29 (m, 1H), 7.29–7.26 (m, 4H), 7.24 (q, J = 1.1 Hz, 1H), 7.12–7.09 (m, 4H), 7.08–7.03 (m, 5H).

^{13}C NMR (100 MHz, Dichloromethane-d2 [54.0 ppm], ppm) δ = 148.1 (2C), 147.9 (1C), 144.7 (1C), 129.8 (4C), 129.0 (1C), 128.6 (1C), 127.2 (2C), 125.0 (4C), 124.6 (1C), 124.2 (2C), 123.7 (2C), 122.8 (1C).

IR (ATR, ṽ) = 3029, 2927, 2847, 1584, 1530, 1492, 1480, 1451, 1429, 1415, 1324, 1315, 1272, 1207, 1191, 1171, 1153, 1137, 1112, 1074, 1050, 1028, 1011, 997, 979, 958, 919, 890, 841, 816, 751, 727, 690, 636, 618, 582, 533 cm^{-1}.

EI-MS (m/z, 70 eV, 70 °C): 328/327 (26/100) [M]$^+$, 77 (10); HRMS–EI (m/z): [M]$^+$ calcd for $C_{22}H_{17}N_1S_1$, 327.1081; found, 327.1083.

4-(2-(3,4-Ethylenedioxy)thienyl)-N,N-bis(4-methoxyphenyl)aniline (14)

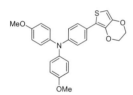

Via CH activation:

4-Bromo-N,N-bis(4-methoxyphenyl)aniline (1) (500 mg, 1.30 mmol, 1.00 equiv), PEPPSI™-iPr (44.3 mg, 65.1 µmol, 0.0500 equiv), P(Cy)$_3$ (52.1 mg, 130 µmol, 0.100 equiv), K$_2$CO$_3$ (270 mg, 1.95 mmol, 1.50 equiv) and pivalic acid (39.9 mg, 390 µmol, 0.300 equiv) were added to a 20 mL crimp cap vial and evacuated and backfilled with argon. EDOT (277 mg, 208 µL, 1.95 mmol, 1.50 equiv) and dry N,N-dimethylformamide (15.0 mL) were added *via* syringe, and the reaction mixture was heated to 100°C under constant stirring. The reaction progress was monitored *via* TLC. After 4 h, the product mixture was cooled to 22 °C, poured into 50 mL water, and extracted with dichloromethane (3x 50 mL). The combined organic phases were washed with water (3x 50 mL) and brine (3x 50 mL), dried over anhydrous Na$_2$SO$_4$, and the solvent was removed under reduced pressure. The crude product was purified *via* column chromatography on silica gel (eluent cyclohexane/ethyl acetate 4:1) to yield 4-(2-(3,4-ethylenedioxy)thienyl)-N,N-bis(4-methoxyphenyl)aniline (264 mg, 593 µmol, 46% yield) as a dark solid.

Via SUZUKI-MIYAURA cross-coupling:

4-Bromo-N,N-bis(4-methoxyphenyl)aniline (1) (1.50 g, 3.90 mmol, 1.00 equiv), 2-pinacolboryl-3,4-ethylenedioxythiophene (16) (2.09 g, 7.81 mmol, 2.00 equiv), Pd(dppf)Cl$_2$ (143 mg, 195 μmol, 0.0500 equiv) and K$_2$CO$_3$ (3.24 g, 23.4 mmol, 6.00 equiv) were added to a 50 mL crimp cap vial and refilled with argon three times. 40 mL of degassed dimethyl sulfoxide was added, and the reaction mixture was heated to 90 °C. The reaction progress was monitored *via* TLC, and after 24 h, no more starting material was observed, and the product mixture was cooled to 22 °C. The mixture was poured into 50 mL water and extracted with ethyl acetate (3x 50 mL). The combined organic phases were washed with water (3x 100 mL) and brine (3x 100 mL), dried over anhydrous Na$_2$SO$_4$, and the solvent was removed under reduced pressure. The crude product was purified *via* column chromatography on silica gel (eluent cyclohexane/ethyl acetate 10:1) to yield 4-(2-(3,4-ethylenedioxy)thienyl)-N,N-bis(4-methoxyphenyl)aniline (1.22 g, 2.73 mmol, 70% yield) as a dark solid.

M.p.: 128 °C.

R$_f$ = 0.2 (cyclohexane/ethyl acetate 10:1).

^{1}H NMR (400 MHz, Dichloromethane-d2 [5.32 ppm], ppm) δ = 7.50–7.48 (m, 2H), 7.06–7.02 (m, 4H), 6.89–6.87 (m, 2H), 6.85–6.81 (m, 4H), 6.20 (s, 1H), 4.27–4.20 (m, 4H), 3.78 (s, 6H).

^{13}C NMR (100 MHz, Dichloromethane-d2 [54.0 ppm], ppm) δ = 156.6 (2C), 148.0 (1C), 142.9 (1C), 141.3 (2C), 137.8 (1C), 127.2 (2C), 127.1 (4C), 125.9 (1C), 120.9 (2C), 118.0 (1C), 115.2 (4C), 96.5 (1C), 65.3 (1C), 65.2 (1C), 56.0 (2C).

IR (ATR, ṽ) = 3108, 3034, 2928, 2836, 1604, 1500, 1489, 1460, 1453, 1436, 1421, 1364, 1317, 1289, 1276, 1264, 1237, 1191, 1177, 1169, 1143, 1109, 1069, 1030, 980, 953, 922, 901, 823, 781, 724, 707, 683, 671, 633, 601, 572, 562 cm^{-1}.

FAB-MS (*m/z*, 3-NBA): 447/446/445 (13/41/100) [M]$^+$; HRMS–FAB (*m/z*): [M]$^+$ calcd for C$_{26}$H$_{23}$O$_4$N$_1$S$_1$, 445.1342, found 445.1344.

2-Pinacolborylthiophene (15)

n-Butyllithium solution (1.90 g, 11.9 mL, 29.7 mmol, 2.50M, 1.00 equiv) was added dropwise to a solution of thiophene (2.50 g, 2.38 mL, 29.7 mmol, 1.00 equiv) in tetrahydrofuran (80.0 mL) at −78 °C. The solution was stirred for 30 min at 22 °C. After cooling to −78 °C, 2-isopropoxy-4,4,5,5-tetramethyl-1,3,2-dioxaborolane (5.53 g, 6.06 mL, 29.7 mmol, 1.00 equiv) was added, and the reaction mixture was stirred for 20 min at 22 °C. After evaporating the solvent under reduced pressure, the crude mixture was purified *via* column chromatography on silica gel (eluent cyclohexane/ethyl acetate 4:1) to yield 2-pinacolborylthiophene (5.43 g, 25.8 mmol, 87% yield) as an off-white solid.

M.p.: 66 °C.

R_f = 0.5 (cyclohexane/ethyl acetate 4:1).

^{1}H NMR (400 MHz, Chloroform-d [7.27 ppm], ppm) δ = 7.68–7.64 (m, 2H), 7.21 (dd, *J* = 3.4 Hz, *J* = 4.7 Hz, 1H), 1.36 (s, 12H).

^{13}C NMR (100 MHz, Chloroform-d [77.0 ppm], ppm) δ = 137.1 (1C), 132.3 (1C), 128.2 (1C), 84.0 (2C), 24.7 (4C), C-B missing due to coupling (1C).

^{11}B NMR (128 MHz, Chloroform-d, ppm) δ = 10.3.

IR (ATR, ṽ) = 2997, 2978, 2928, 1519, 1482, 1468, 1422, 1361, 1329, 1295, 1266, 1208, 1166, 1135, 1112, 1078, 1055, 1017, 1000, 956, 929, 914, 849, 826, 778, 754, 722, 687, 664, 630, 578 cm^{-1}.

EI-MS (*m/z*, 70 eV, 110 °C): 210 (76) [M]$^+$, 195 (60), 181 (31), 131 (32), 124 (83), 111/110 (100,52), 69 (74); HRMS–EI (*m/z*): [M]$^+$ calcd for $C_{10}H_{15}O_2B_1S_1$, 210.0880; found, 210.0882.

2-Pinacolboryl-3,4-ethylenedioxythiophene (16)

n-Butyllithium solution (1.82 g, 11.4 mL, 28.4 mmol, 2.50M, 1.01 equiv) was added dropwise to a solution of EDOT (4.00 g, 2.99 mL, 28.1 mmol, 1.00 equiv) in tetrahydrofuran (100 mL) at −78 °C. The solution was stirred for 30 min at 22 °C. After cooling to −78 °C, 2-isopropoxy-4,4,5,5-tetramethyl-1,3,2-dioxaborolane (5.23 g, 5.74 mL, 28.1 mmol, 1.00 equiv) was added, and the reaction mixture was stirred for 4 h at 22 °C, reaction progress was monitored *via* TLC. The solvent was removed under reduced pressure; the solid residue was taken up in dichloromethane (100 mL). 100 mL of 6N HCl was added, the phases were separated, and the organic phase was dried over anhydrous Na_2SO_4. After evaporating the solvent under reduced pressure, the crude product was recrystallized from *n*-hexane to yield 2-pinacolboryl-3,4-ethylenedioxythiophene (6.50 g, 24.2 mmol, 86% yield) as off-white crystals.

M.p.: 91 °C.

R_f = 0.2 (cyclohexane/ethyl acetate 3:1)

^{1}H NMR (400 MHz, Chloroform-d [7.27 ppm], ppm) δ = 6.63 (s, 1H), 4.32–4.30 (m, 2H), 4.20–4.18 (m, 2H), 1.34 (s, 12H).

^{13}C NMR (100 MHz, Chloroform-d [77.0 ppm], ppm) δ = 149.0 (1C), 142.3 (1C), 107.5 (1C), 83.8 (2C), 65.0 (1C), 64.3 (1C), 24.7 (4C), C-B missing due to coupling (1C).

^{11}B NMR (128 MHz, Chloroform-d [77.0 ppm], ppm) δ = 9.7.

IR (ATR, $\tilde{v}$) = 3105, 2979, 2932, 2874, 1577, 1486, 1476, 1438, 1387, 1371, 1364, 1336, 1309, 1269, 1247, 1211, 1181, 1169, 1139, 1115, 1067, 1047, 1007, 983, 965, 941, 905, 861, 851, 839, 816, 756, 738, 704, 664, 646, 577, 564 cm^{-1}.

EI-MS (*m/z*, 70 eV, 50 °C): 268 (100) [M]$^+$, 225 (67), 181 (52), 169/168 (50/52), 131 (29); HRMS–EI (*m/z*): [M]$^+$ calcd for $C_{12}H_{17}O_4B_1S_1$, 268.0935; found, 268.0937.

2-Pinacolborylselenophene (17)

n-Butyllithium solution (1.23 g, 7.71 mL, 19.3 mmol, 2.50M, 1.01 equiv) was added dropwise to a solution of selenophene (2.50 g, 1.64 mL, 19.1 mmol, 1.00 equiv) in tetrahydrofuran (50.0 mL) at −78 °C. The solution was stirred for 30 min at 22 °C. After cooling to −78 °C, 2-isopropoxy-4,4,5,5-tetramethyl-1,3,2-dioxaborolane (3.55 g, 3.89 mL, 19.1 mmol, 1.00 equiv) was added, and the reaction mixture was stirred for 20 min at 22 °C. After evaporating the solvent under reduced pressure, the crude mixture was purified *via* column chromatography on silica gel (eluent cyclohexane/ethyl acetate 4:1 → 2:1) to yield 2-pinacolborylselenophene (3.81 g, 14.8 mmol, 78% yield) as a yellowish solid.

M.p.: 55 °C.

R_f = 0.75 (cyclohexane/ethyl acetate 4:1).

^{1}H NMR (400 MHz, Chloroform-d [7.27 ppm], ppm) δ = 8.38 (d, J = 5.1 Hz, 1H), 7.96 (d, J = 3.6 Hz, 1H), 7.47 (dd, J = 3.8 Hz, J = 4.8 Hz, 1H), 1.36 (s, 12H).

^{13}C NMR (100 MHz, Chloroform-d [77.0 ppm], ppm) δ = 139.6 (1C), 137.7 (1C), 131.0 (1C), 84.1 (2C), 24.7 (4C), C-B missing due to coupling (1C).

^{11}B NMR (128 MHz, Chloroform-d, ppm) δ = 29.7.

IR (ATR, ṽ) = 2996, 2976, 2929, 1531, 1468, 1452, 1434, 1390, 1371, 1349, 1327, 1298, 1266, 1208, 1166, 1135, 1111, 1079, 1037, 1007, 986, 950, 928, 914, 849, 827, 807, 752, 708, 683, 660, 645, 577 cm^{-1}.

EI-MS (*m/z*, 70 eV, 20 °C): 258/256 (100/46) [M]$^+$, 198/196 (67/32), 172 (45), 159/158/157/156 (73/46/42/32); HRMS–EI (*m/z*): [M]$^+$ calcd for $C_{10}H_{15}O_2B_1Se_1$, 258.0325; found, 258.0324.

4-(2-Thienyl)-*N,N*-bis(3-methoxyphenyl)aniline (18)

4-Bromo-*N,N*-bis(3-methoxyphenyl)aniline (2) (1000 mg, 2.60 mmol, 1.00 equiv), 2-pinacolborylthiophene (15) (656 mg, 3.12 mmol, 1.20 equiv), Pd(dppf)Cl$_2$ (95.2 mg, 130 µmol, 0.0500 equiv) and K$_2$CO$_3$ (2.16 g, 15.6 mmol, 6.00 equiv) were added to a 50 mL crimp cap vial and refilled with argon three times. 30 mL of degassed dimethyl sulfoxide was added, and the reaction mixture was heated to 90 °C. The reaction progress was monitored *via* TLC; after 2 h, the product mixture was cooled to 22 °C. The product mixture was poured into 100 mL water and extracted with ethyl acetate (3x 100 mL). The combined organic phases were washed with water (3x 200 mL) and brine (3x 200 mL), dried over anhydrous Na$_2$SO$_4$, and the solvents were removed under reduced pressure. The crude product was purified *via* column chromatography on silica gel (eluent cyclohexane/ethyl acetate 10:1 + 1% triethylamine) to yield 4-(2-thienyl)-*N,N*-bis(3-methoxyphenyl)aniline (992 mg, 2.56 mmol, 98% yield) as a red solid.

M.p.: 87 °C.

R$_f$ = 0.5 (cyclohexane/ethyl acetate 4:1).

^{1}H NMR (400 MHz, Dichloromethane-d2 [5.32 ppm], ppm) δ = 7.50 (d, *J* = 8.6 Hz, 2H), 7.26–7.24 (m, 2H), 7.18 (t, *J* = 8.1 Hz, 2H), 7.08–7.06 (m, 3H), 6.69–6.59 (m, 6H), 3.72 (s, 6H).

^{13}C NMR (100 MHz, Dichloromethane-d2 [54.0 ppm], ppm) δ = 161.2 (2C), 149.2 (2C), 147.6 (1C), 144.7 (1C), 130.4 (2C), 129.3 (1C), 128.6 (1C), 127.1 (2C), 124.6 (3C), 122.9 (1C), 117.5 (2C), 110.9 (2C), 109.1 (2C), 55.8 (2C).

IR (ATR, ṽ) = 3064, 2999, 2951, 2934, 2912, 2904, 2832, 1587, 1575, 1561, 1531, 1483, 1443, 1435, 1415, 1339, 1317, 1300, 1264, 1248, 1205, 1174, 1166, 1154, 1140, 1115, 1095, 1082, 1041, 993, 977, 963, 958, 858, 844, 826, 788, 779, 761, 728, 714, 687, 643, 628, 584, 565, 535 cm^{-1}.

FAB-MS (*m/z*, 3-NBA): 388/387 (51/100) [M]$^+$; HRMS–FAB (*m/z*): [M]$^+$ calcd for C$_{24}$H$_{21}$O$_2$N$_1$S$_1$, 387.1288, found 387.1287.

4-(2-Thienyl)-*N,N*-bis(2-methoxyphenyl)aniline (19)

4-Bromo-*N,N*-bis(2-methoxyphenyl)aniline (3) (1.00 g, 2.60 mmol, 1.00 equiv), 2-pinacolborylthiophene (15) (656 mg, 3.12 mmol, 1.20 equiv), Pd(dppf)Cl$_2$ (95.2 mg, 130 μmol, 0.0500 equiv) and K$_2$CO$_3$ (2.16 g, 15.6 mmol, 6.00 equiv) were added to a 50 mL crimp cap vial and refilled with argon three times. 35 mL of degassed dimethyl sulfoxide was added, and the reaction mixture was heated to 90 °C. The reaction progress was monitored *via* TLC; after 20 h, the product mixture was cooled to 22 °C. The mixture was poured into 100 mL water and extracted with ethyl acetate (3x 100 mL). The combined organic phases were washed with water (3x 200 mL) and brine (3x 200 mL), dried over anhydrous Na$_2$SO$_4$, and the solvent was removed under reduced pressure. The crude product was purified *via* column chromatography on silica gel (eluent cyclohexane/ethyl acetate 15:1) to yield 4-(2-thienyl)-*N,N*-bis(2-methoxyphenyl)aniline (884 mg, 2.28 mmol, 88% yield) as a red solid.

M.p.: 89 °C.

R$_f$ = 0.5 (cyclohexane/ethyl acetate 6:1).

^{1}H NMR (400 MHz, Dichloromethane-d2 [5.32 ppm], ppm) δ = 7.35 (d, *J* = 8.8 Hz, 2H), 7.23–7.16 (m, 6H), 7.04–6.99 (m, 3H), 6.94 (t, *J* = 7.6 Hz, 2H), 6.53 (d, *J* = 8.6 Hz, 2H), 3.73 (s, 6H).

^{13}C NMR (100 MHz, Dichloromethane-d2 [54.0 ppm], ppm) δ = 156.3 (2C), 148.5 (1C), 145.6 (1C), 135.3 (2C), 129.8 (2C), 128.4 (1C), 127.3 (2C), 126.6 (2C), 125.3 (1C), 123.5 (1C), 121.8 (2C), 121.6 (1C), 116.7 (2C), 113.4 (2C), 56.2 (2C).

IR (ATR, ṽ) = 3407, 3394, 3111, 3023, 3000, 2952, 2924, 2873, 2851, 2832, 1608, 1589, 1536, 1494, 1455, 1432, 1402, 1332, 1322, 1293, 1268, 1238, 1207, 1181, 1156, 1111, 1081, 1044, 1024, 1001, 975, 956, 936, 847, 827, 815, 798, 749, 694, 659, 649, 637, 620, 582, 567, 547, 540 cm^{-1}.

FAB-MS (*m/z*, 3-NBA): 388/387 (50/100) [M]$^+$, 97 (32), 95 (43); HRMS–FAB (*m/z*): [M]$^+$ calcd for C$_{24}$H$_{21}$O$_2$N$_1$S$_1$, 387.1288, found 387.1288.

4-(2-Thienyl)-*N*,*N*-bis(2,4-dimethoxyphenyl)aniline (20)

4-Bromo-*N*,*N*-bis(2,4-dimethoxyphenyl)aniline (4) (1.00 g, 2.25 mmol, 1.00 equiv), 2-pinacolborylthiophene (15) (567 mg, 2.70 mmol, 1.20 equiv), Pd(dppf)Cl$_2$ (82.3 mg, 113 μmol, 0.0500 equiv) and K$_2$CO$_3$ (1.87 g, 13.5 mmol, 6.00 equiv) were added to a 50 mL crimp cap vial and refilled with argon three times. 30 mL of degassed dimethyl sulfoxide was added, and the reaction mixture was heated to 90 °C. The reaction progress was monitored *via* TLC, and after 2 h, the product mixture was cooled to 22 °C. The mixture was poured into 100 mL water and extracted with ethyl acetate (3x 100 mL). The combined organic phases were washed with water (3x 200 mL) and brine (3x 200 mL), dried over anhydrous Na$_2$SO$_4$, and the solvents were removed under reduced pressure. The crude product was purified *via* column chromatography on silica gel (eluent cyclohexane/ethyl acetate 10:1 + 1% tri-ethylamine) to yield 4-(2-thienyl)-*N*,*N*-bis(2,4-dimethoxyphenyl)aniline (719 mg, 1.61 mmol, 71% yield) as a brown solid.

M.p.: 170 °C.

R$_f$ = 0.55 (cyclohexane/ethyl acetate 4:1).

^{1}H NMR (400 MHz, Dichloromethane-d2 [5.32 ppm], ppm) δ = 7.31 (d, *J* = 8.6 Hz, 2H), 7.25 (d, *J* = 8.6 Hz, 2H), 7.15–7.12 (m, 2H), 7.01 (dd, *J* = 3.8 Hz, *J* = 4.9 Hz, 1H), 6.57 (d, *J* = 2.5 Hz, 2H), 6.48 (dd, *J* = 2.6 Hz, *J* = 8.5 Hz, 2H), 6.40 (d, *J* = 8.8 Hz, 2H), 3.81 (s, 6H), 3.77 (s, 6H).

^{13}C NMR (100 MHz, Dichloromethane-d2 [54.0 ppm], ppm) δ = 159.8 (2C), 157.6 (2C), 149.5 (1C), 145.8 (1C), 131.1 (2C), 128.4 (1C), 128.0 (2C), 126.7 (2C), 124.0 (1C), 123.2 (1C), 121.3 (1C), 114.6 (2C), 105.6 (2C), 100.4 (2C), 56.2 (2C), 56.0 (2C).

IR (ATR, ṽ) = 3108, 3006, 2961, 2934, 2874, 2853, 2836, 1605, 1579, 1537, 1500, 1460, 1436, 1417, 1343, 1296, 1275, 1255, 1238, 1208, 1187, 1156, 1136, 1113, 1081, 1028, 959, 948, 931, 901, 880, 833, 820, 803, 742, 724, 697, 636, 605, 585, 564 cm^{-1}.

FAB-MS (*m/z*, 3-NBA): 448/447 (42/100) [M]$^+$, 133 (43), 95 (37), 91 (33); HRMS–FAB (*m/z*): [M]$^+$ calcd for C$_{26}$H$_{25}$O$_4$N$_1$S$_1$, 447.1499, found 447.1500.

4-(2-Thienyl)-*N*,*N*-bis(3,4,5-trimethoxyphenyl)aniline (21)

4-Bromo-*N*,*N*-bis(3,4,5-trimethoxyphenyl)aniline (5) (1.00 g, 1.98 mmol, 1.00 equiv), 2-pinacolborylthiophene (15) (500 mg, 2.38 mmol, 1.20 equiv), Pd(dppf)Cl$_2$ (72.5 mg, 99.1 µmol, 0.0500 equiv) and K$_2$CO$_3$ (1.64 g, 11.9 mmol, 6.00 equiv) were added to a 50 mL crimp cap vial and refilled with argon three times. 20 mL of degassed dimethyl sulfoxide was added, and the reaction mixture was heated to 90 °C. The reaction progress was monitored *via* TLC, and after 3 h, the product mixture was cooled to 22 °C. The mixture was poured into 100 mL water and extracted with ethyl acetate (3x 100 mL). The combined organic phases were washed with water (3x 200 mL) and brine (3x 200 mL), dried over anhydrous Na$_2$SO$_4$, and the solvent was removed under reduced pressure. The crude product was purified *via* column chromatography on silica gel (eluent cyclohexane/ethyl acetate 6:1 → 2:1 + 1% triethylamine) to yield 4-(2-thienyl)-*N*,*N*-bis(3,4,5-trimethoxyphenyl)aniline (991 mg, 1.95 mmol, 98% yield) as a grey solid.

M.p.: 163 °C.

R_f = 0.37 (cyclohexane/ethyl acetate 4:1).

^{1}H NMR (400 MHz, Dichloromethane-d2 [5.32 ppm], ppm) δ = 7.48 (d, *J* = 8.6 Hz, 2H), 7.25–7.23 (m, 2H), 7.07–7.02 (m, 3H), 6.35 (s, 4H), 3.77 (s, 6H), 3.70 (s, 12H).

^{13}C NMR (100 MHz, Dichloromethane-d2 [54.0 ppm], ppm) δ = 154.4 (4C), 147.9 (1C), 144.8 (1C), 143.7 (2C), 135.2 (2C), 128.6 (1C), 128.3 (1C), 126.9 (2C), 124.4 (1C), 123.2 (2C), 122.6 (1C), 103.2 (4C), 61.1 (2C), 56.6 (4C).

IR (ATR, ṽ) = 3087, 3009, 2961, 2932, 2918, 2881, 2837, 2825, 1581, 1560, 1531, 1496, 1462, 1445, 1422, 1408, 1361, 1344, 1307, 1266, 1228, 1207, 1181, 1122, 1081, 1051, 1023, 1001, 958, 921, 907, 846, 824, 807, 793, 769, 751, 731, 701, 662, 646, 637, 622, 603, 588, 535, 523 cm^{-1}.

FAB-MS (*m/z*, 3-NBA): 664/663/662 (39/100/47) [unknown impurity, visible in scans before and after product], 648/647 (27/58) [unknown impurity, visible in scans before and after product], 508/507 (41/80) [M]$^+$, 441 (40), 154 (54), 147

(31), 137/136 (31/41), 97/95/91 (33/45/41); HRMS–FAB (*m/z*): [M]$^+$ calcd for C$_{28}$H$_{29}$O$_6$N$_1$S$_1$, 507.1710, found 507.1709.

4-(2-Thienyl)-*N,N*-bis(4-*tert*-butylphenyl)aniline (22)

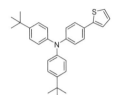

4-Bromo-*N,N*-bis(4-*tert*-butylphenyl)aniline (6) (1.00 g, 2.29 mmol, 1.00 equiv), 2-pinacolborylthiophene (15) (578 mg, 2.75 mmol, 1.20 equiv), Pd(dppf)Cl2 (83.8 mg, 115 µmol, 0.0500 equiv) and K$_2$CO$_3$ (1.90 g, 13.7 mmol, 6.00 equiv) were added to a 50 mL crimp cap vial and refilled with argon three times. 35 mL of degassed dimethyl sulfoxide was added, and the reaction mixture was heated to 90 °C. The reaction progress was monitored *via* TLC, and after 19.5 h, the product mixture was cooled to 22 °C. The mixture was poured into 100 mL water and extracted with ethyl acetate (3x 100 mL). The combined organic phases were washed with water (3x 200 mL) and brine (3x 200 mL), dried over anhydrous Na$_2$SO$_4$, and the solvent was removed under reduced pressure. The crude product was purified *via* column chromatography on silica gel (eluent cyclohexane/ethyl acetate 10:1) to yield 4-(2-thienyl)-*N,N*-bis(4-*tert*-butylphenyl)aniline (763 mg, 1.74 mmol, 76% yield) as an off-white solid.

M.p.: 204 °C.

R$_f$ = 0.8 (cyclohexane/ethyl acetate 10:1).

^{1}H NMR (400 MHz, Dichloromethane-d2 [5.32 ppm], ppm) δ = 7.47–7.44 (m, 2H), 7.32–7.28 (m, 4H), 7.22 (d, *J* = 4.3 Hz, 2H), 7.07–6.99 (m, 7H), 1.32 (s, 18H).

^{13}C NMR (100 MHz, Dichloromethane-d2 [54.0 ppm], ppm) δ = 148.3 (1C), 146.7 (2C), 145.4 (2C), 144.9 (1C), 128.5 (1C), 128.1 (1C), 127.0 (2C), 126.7 (4C), 124.8 (4C), 124.3 (1C), 123.1 (2C), 122.5 (1C), 34.8 (2C), 31.7 (6C).

IR (ATR, ṽ) = 3037, 2958, 2900, 2861, 1598, 1533, 1507, 1496, 1460, 1431, 1391, 1361, 1320, 1298, 1278, 1266, 1210, 1200, 1181, 1111, 1082, 1050, 1021, 1014, 956, 945, 922, 849, 829, 817, 751, 742, 737, 724, 696, 579, 568, 545 cm^{-1}.

FAB-MS (*m/z*, 3-NBA): 440/439 (49/100) [M]$^+$, 154 (59), 137/136 (37/41); HRMS–FAB (*m/z*): [M]$^+$ calcd for C$_{30}$H$_{33}$N$_1$S$_1$, 439.2328, found 439.2329.

4-(2-Thienyl)-*N*,*N*-bis(4-methylthiophenyl)aniline (23)

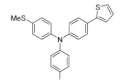

4-Bromo-*N*,*N*-bis(4-methylthiophenyl)aniline (7) (1.00 g, 2.40 mmol, 1.00 equiv), 2-pinacolborylthiophene (15) (605 mg, 2.88 mmol, 1.20 equiv), Pd(dppf)Cl$_2$ (87.9 mg, 120 µmol, 0.0500 equiv) and K$_2$CO$_3$ (1.99 g, 14.4 mmol, 6.00 equiv) were added to a 50 mL crimp cap vial and refilled with argon three times. 24 mL of degassed dimethyl sulfoxide was added, and the reaction mixture was heated to 90 °C. The reaction progress was monitored *via* TLC, and after 3 h, the product mixture was cooled to 22 °C. The mixture was poured into 100 mL water and extracted with ethyl acetate (3x 100 mL). The combined organic phases were washed with water (3x 200 mL) and brine (3x 200 mL), dried over anhydrous Na$_2$SO$_4$, and the solvents were removed under reduced pressure. The crude product was purified *via* column chromatography on silica gel (eluent cyclohexane/ethyl acetate 10:1 + 1% tri-ethylamine) to yield 4-(2-thienyl)-*N*,*N*-bis(4-methylthiophenyl)aniline (977 mg, 2.33 mmol, 97% yield) as a red solid.

M.p.: 106 °C.

R_f = 0.9 (cyclohexane/ethyl acetate 4:1).

^{1}H NMR (400 MHz, Dichloromethane-d2 [5.32 ppm], ppm) δ = 7.48 (d, *J* = 8.5 Hz, 2H), 7.24 (d, *J* = 4.4 Hz, 2H), 7.19 (d, *J* = 8.6 Hz, 4H), 7.07–7.02 (m, 7H), 2.47 (s, 6H).

^{13}C NMR (100 MHz, Dichloromethane-d2 [54.0 ppm], ppm) δ = 147.5 (1C), 145.4 (2C), 144.6 (1C), 133.0 (2C), 129.1 (1C), 128.8 (4C), 128.6 (1C), 127.2 (2C), 125.5 (4C), 124.6 (1C), 123.9 (2C), 122.8 (1C), 17.0 (2C).

IR (ATR, ṽ) = 3024, 2983, 2917, 2854, 1900, 1606, 1585, 1568, 1530, 1486, 1429, 1422, 1400, 1358, 1322, 1307, 1282, 1268, 1197, 1183, 1173, 1130, 1111, 1091, 1058, 1010, 958, 916, 847, 820, 812, 755, 722, 683, 637, 582, 535, 524 cm^{-1}.

FAB-MS (*m/z*, 3-NBA): 421/420/419 (20/37/100) [M]$^+$; HRMS–FAB (*m/z*): [M]$^+$ calcd for C$_{24}$H$_{21}$N$_1$S$_3$, 419.0831, found 419.0828.

9-(4-(2-Thienyl)phenyl)carbazole (24)

9-(4-Bromophenyl)carbazole (1.00 g, 3.10 mmol, 1.00 equiv), 2-pinacolborylthiophene (15) (783 mg, 3.72 mmol, 1.20 equiv), Pd(dppf)Cl$_2$ (114 mg, 155 µmol, 0.0500 equiv) and K$_2$CO$_3$ (2.57 g, 18.6 mmol, 6.00 equiv) were added to a 50 mL crimp cap vial and refilled with argon three times. 30 mL of degassed dimethyl sulfoxide was added, and the reaction mixture was heated to 90 °C. The reaction progress was monitored *via* TLC, and after 1.7 h, the product mixture was cooled to 22 °C. The mixture was poured into 100 mL water and extracted with ethyl acetate (3x 100 mL). The combined organic phases were washed with water (3x 200 mL) and brine (3x 200 mL), dried over anhydrous Na$_2$SO$_4$, and the solvent was removed under reduced pressure. The crude product was purified *via* column chromatography on silica gel (eluent cyclohexane/ethyl acetate 100:1) to yield 9-(4-(2-thienyl)phenyl)carbazole (755 mg, 2.32 mmol, 75% yield) as an off-white solid.

M.p.: 185 °C.

R$_f$ = 0.8 (cyclohexane/ethyl acetate 4:1)

^{1}H NMR (400 MHz, Dichloromethane-d2 [5.32 ppm], ppm) δ = 8.17 (d, *J* = 7.6 Hz, 2H), 7.89–7.85 (m, 2H), 7.62–7.59 (m, 2H), 7.49–7.42 (m, 5H), 7.38 (dd, *J* = 0.8 Hz, *J* = 5.1 Hz, 1H), 7.33–7.29 (m, 2H), 7.16 (dd, *J* = 3.7 Hz, *J* = 4.9 Hz, 1H).

^{13}C NMR (100 MHz, Dichloromethane-d2 [54.0 ppm], ppm) δ = 143.9 (1C), 141.3 (2C), 137.3 (1C), 134.0 (1C), 128.8 (1C), 127.9 (2C), 127.7 (2C), 126.6 (2C), 125.9 (1C), 124.2 (1C), 123.9 (2C), 120.8 (2C), 120.6 (2C), 110.3 (2C).

IR (ATR, ṽ) = 3094, 3063, 3050, 3040, 2924, 1938, 1904, 1870, 1783, 1625, 1596, 1564, 1536, 1503, 1477, 1449, 1431, 1415, 1364, 1350, 1334, 1316, 1299, 1276, 1259, 1228, 1214, 1187, 1169, 1145, 1118, 1108, 1082, 1055, 1018, 996, 958, 939, 931, 914, 856, 847, 839, 822, 759, 747, 724, 703, 677, 642, 623, 585, 568, 533 cm^{-1}.

EI-MS (*m/z*, 70 eV, 70 °C): 327/326/325 (7/24/100) [M]$^+$; HRMS–EI (*m/z*): [M]$^+$ calcd for C$_{22}$H$_{15}$N$_1$S$_1$, 325.0925; found, 325.0897.

9-(4-(2-Thienyl)phenyl)-3,6-dimethoxycarbazole (25)

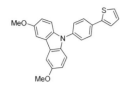

9-(4-Bromophenyl)-3,6-dimethoxycarbazole (9) (1.00 g, 2.62 mmol, 1.00 equiv), 2-pinacolborylthiophene (15) (605 mg, 2.88 mmol, 1.10 equiv), Pd(dppf)Cl$_2$ (95.7 mg, 131 μmol, 0.0500 equiv) and K$_2$CO$_3$ (2.17 g, 15.7 mmol, 6.00 equiv) were added to a 50 mL crimp cap vial and refilled with argon three times. 30 mL of degassed dimethyl sulfoxide was added, and the reaction mixture was heated to 90 °C. The reaction progress was monitored *via* TLC, and after 1 h, the product mixture was cooled to 22 °C. The mixture was poured into 100 mL water and extracted with ethyl acetate (3x 100 mL). The combined organic phases were washed with water (3x 200 mL) and brine (3x 200 mL), dried over anhydrous Na$_2$SO$_4$, and the solvent was removed under reduced pressure. The crude product was purified *via* column chromatography on silica gel (eluent cyclohexane/ethyl acetate 20:1 → 10:1) to yield 9-(4-(2-thienyl)phenyl)-3,6-dimethoxycarbazole (940 mg, 2.44 mmol, 93% yield) as an off-white solid.

M.p.: 133 °C.

R_f = 0.52 (cyclohexane/ethyl acetate 4:1).

^{1}H NMR (400 MHz, Dichloromethane-d2 [5.32 ppm], ppm) δ = 7.85–7.82 (m, 2H), 7.59–7.56 (m, 4H), 7.43 (dd, J = 1.1 Hz, J = 3.7 Hz, 1H), 7.39–7.36 (m, 3H), 7.15 (dd, J = 3.5 Hz, J = 5.1 Hz, 1H), 7.04 (dd, J = 2.7 Hz, J = 9.0 Hz, 2H), 3.93 (s, 6H).

^{13}C NMR (100 MHz, Dichloromethane-d2 [54.0 ppm], ppm) δ = 154.8 (2C), 144.0 (1C), 137.9 (1C), 136.6 (2C), 133.5 (1C), 128.8 (1C), 127.7 (2C), 127.4 (2C), 125.8 (1C), 124.3 (1C), 124.1 (2C), 115.7 (2C), 111.3 (2C), 103.3 (2C), 56.5 (2C).

IR (ATR, ṽ) = 3054, 2938, 2912, 2894, 2837, 1606, 1581, 1533, 1502, 1487, 1460, 1436, 1412, 1361, 1351, 1329, 1306, 1283, 1242, 1204, 1176, 1159, 1137, 1113, 1084, 1054, 1033, 969, 959, 946, 911, 887, 874, 849, 841, 822, 786, 755, 722, 694, 654, 640, 630, 611, 592, 582, 558 cm^{-1}.

FAB-MS (*m/z*, 3-NBA): 387/386/385 (13/43/100) [M]$^+$, 370 (14); HRMS–FAB (*m/z*): [M]$^+$ calcd for C$_{24}$H$_{19}$O$_2$N$_1$S$_1$, 385.1131, found 385.1129.

9-(4-(2-Thienyl)phenyl)-3,6-di*tert*-butylcarbazole (26)

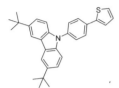 9-(4-Bromophenyl)-3,6-di*tert*-butylcarbazole (10) (1.00 g, 2.30 mmol, 1.00 equiv), 2-pinacolborylthiophene (15) (580 mg, 2.76 mmol, 1.20 equiv), Pd(dppf)Cl$_2$ (84.2 mg, 115 μmol, 0.0500 equiv) and KOH (1.91 g, 13.8 mmol, 6.00 equiv) were added to a 50 mL crimp cap vial and refilled with argon three times. 30 mL of degassed dimethyl sulfoxide was added, and the reaction mixture was heated to 90 °C. The reaction progress was monitored *via* TLC, and after 1.5 h, the product mixture was cooled to 22 °C. The mixture was poured into 100 mL water and extracted with ethyl acetate (3x 100 mL). The combined organic phases were washed with water (3x 200 mL) and brine (3x 200 mL), dried over anhydrous Na$_2$SO$_4$, and the solvent was removed under reduced pressure. The crude product was purified *via* column chromatography on silica gel (eluent cyclohexane/ethyl acetate 100:1) to yield 9-(4-(2-thienyl)phenyl)-3,6-di*tert*-butylcarbazole (817 mg, 1.87 mmol, 81% yield) as an off-white solid.

M.p.: 193 °C.

R$_f$ = 0.85 (cyclohexane/ethyl acetate 20:1).

^{1}H NMR (400 MHz, Dichloromethane-d2 [5.32 ppm], ppm) δ = 8.18–8.17 (m, 2H), 7.86–7.83 (m, 2H), 7.61–7.58 (m, 2H), 7.51–7.48 (m, 2H), 7.44 (dd, J = 1.1 Hz, J = 3.7 Hz, 1H), 7.42–7.39 (m, 2H), 7.37 (dd, J = 1.1 Hz, J = 5.2 Hz, 1H), 7.15 (dd, J = 3.5 Hz, J = 5.1 Hz, 1H), 1.47 (s, 18H).

^{13}C NMR (100 MHz, Dichloromethane-d2 [54.0 ppm], ppm) δ = 144.0 (1C), 143.6 (2C), 139.6 (2C), 137.9 (1C), 133.5 (1C), 128.8 (1C), 127.7 (2C), 127.5 (2C), 125.8 (1C), 124.3 (2C), 124.1 (1C), 123.9 (2C), 116.9 (2C), 109.8 (2C), 35.2 (2C), 32.3 (6C).

IR (ATR, ṽ) = 3043, 2956, 2901, 2864, 1626, 1604, 1536, 1503, 1487, 1470, 1431, 1409, 1391, 1363, 1326, 1293, 1258, 1238, 1207, 1174, 1145, 1129, 1105, 1079, 1050, 1035, 1017, 958, 907, 880, 850, 841, 813, 732, 694, 652, 640, 628, 611, 599, 535 cm^{-1}.

FAB-MS (*m/z*, 3-NBA): 439/438/437 (15/46/100) [M]$^+$, 423/422 (9/27), 382 (10), 154 (21), 137/136 (12/19); HRMS–FAB (*m/z*): [M]$^+$ calcd for $C_{30}H_{31}N_1{}^{32}S_1$, 437.2172, found 437.2171.

4-(2-(3,4-Ethylenedioxy)thienyl)-*N*,*N*-bis(3,4,5-trimethoxyphenyl)aniline (27)

4-Bromo-*N*,*N*-bis(3,4,5-trimethoxyphenyl)aniline (5) (1.00 g, 1.98 mmol, 1.00 equiv), 2-pinacolboryl-3,4-ethylenedioxythiophene (16) (638 mg, 2.38 mmol, 1.20 equiv), Pd(dppf)Cl$_2$ (72.5 mg, 99.1 μmol, 0.0500 equiv) and K$_2$CO$_3$ (1.64 g, 11.9 mmol, 6.00 equiv) were added to a 50 mL crimp cap vial and refilled with argon three times. 25 mL of degassed dimethyl sulfoxide was added, and the reaction mixture was heated to 90 °C. After 18 h, the temperature was increased to 100 °C. The reaction progress was monitored *via* TLC, and after 28 h, the product mixture was cooled to 22 °C. The mixture was poured into 50 mL water and extracted with ethyl acetate (3x 50 mL). The combined organic phases were washed with water (3x 100 mL) and brine (3x 100 mL), dried over anhydrous Na$_2$SO$_{4,}$ and the solvents were removed under reduced pressure. The crude product was purified *via* column chromatography on silica gel (eluent cyclohexane/ethyl acetate 2:1) to yield 4-(2-(3,4-ethylenedioxy)thienyl)-*N*,*N*-bis(3,4,5-trimethoxyphenyl)aniline (376 mg, 665 μmol, 34% yield) as a dark yellow solid.

M.p.: 177 °C.

R_f = 0.15 (cyclohexane/ethyl acetate 2:1).

^{1}H NMR (400 MHz, Dichloromethane-d2 [5.32 ppm], ppm) δ = 7.57 (d, *J* = 8.6 Hz, 2H), 7.01 (d, *J* = 8.8 Hz, 2H), 6.34 (s, 4H), 6.24 (s, 1H), 4.25 (td, *J* = 2.8 Hz, *J* = 4.3 Hz, 4H), 3.76 (s, 6H), 3.69 (s, 12H).

^{13}C NMR (100 MHz, Dichloromethane-d2 [54.0 ppm], ppm) δ = 154.3 (4C), 146.9 (1C), 143.8 (2C), 143.0 (1C), 138.2 (1C), 135.0 (2C), 127.3 (1C), 127.1 (2C), 123.2 (2C), 117.7 (1C), 103.1 (4C), 97.0 (1C), 65.4 (1C), 65.1 (1C), 61.1 (2C), 56.6 (4C).

IR (ATR, ṽ) = 3108, 2927, 2867, 2851, 2839, 2825, 1738, 1584, 1496, 1466, 1451, 1431, 1408, 1361, 1346, 1312, 1272, 1227, 1207, 1179, 1147, 1115, 1074, 1050, 1020, 1001, 986, 969, 921, 905, 840, 830, 817, 807, 792, 766, 754, 731, 714, 698, 690, 652, 615 cm^{-1}.

FAB-MS (*m/z*, 3-NBA): 663 (60), 566/565 (54/100) [M]$^{+}$, 154 (54), 137/136 (37/45), 97 (50), 95 (50), 91 (36); HRMS–FAB (*m/z*): [M]$^{+}$ calcd for $C_{30}H_{31}N_1S_1$, 565.1765, found 565.1766.

4-(2-(3,4-Ethylenedioxy)thienyl)-*N,N*-bis(4-*tert*-butylphenyl)aniline (28)

4-Bromo-*N,N*-bis(4-*tert*-butylphenyl)aniline (6) (1.00 g, 2.29 mmol, 1.00 equiv), 2-pinacolboryl-3,4-ethylene-dioxythiophene (16) (676 mg, 2.52 mmol, 1.10 equiv), Pd(dppf)Cl$_2$ (83.8 mg, 115 μmol, 0.0500 equiv) and K$_2$CO$_3$ (1.90 g, 13.7 mmol, 6.00 equiv) were added to a 50 mL crimp cap vial and refilled with argon three times. 30 mL of degassed dimethyl sulfoxide was added, and the reaction mixture was heated to 90 °C. The reaction progress was monitored *via* TLC, and after 14 h, the product mixture was cooled to 22 °C. The mixture was poured into 50 mL water and extracted with ethyl acetate (3x 50 mL). The combined organic phases were washed with water (3x 100 mL) and brine (3x 100 mL), dried over anhydrous Na$_2$SO$_4$, and the solvent was removed under reduced pressure. The crude product was purified *via* column chromatography on silica gel (eluent cyclohexane/dichloromethane 3:2 + 1% triethyamine) to yield 4-(2-(3,4-ethylenedioxy)thienyl)-*N,N*-bis(4-*tert*-butylphenyl)aniline (415 mg, 834 μmol, 36% yield) as an off-white solid.

M.p.: 189 °C.

R$_f$ = 0.17 (cyclohexane/dichloromethane 2:3).

^{1}H NMR (400 MHz, Dichloromethane-d2 [5.32 ppm], ppm) δ = 7.56–7.52 (m, 2H), 7.30–7.27 (m, 4H), 7.03–6.98 (m, 6H), 6.23 (s, 1H), 4.28–4.21 (m, 4H), 1.31 (s, 18H).

^{13}C NMR (100 MHz, Dichloromethane-d2 [54.0 ppm], ppm) δ = 147.3 (1C), 146.5 (2C), 145.5 (2C), 143.0 (1C), 138.1 (1C), 127.2 (2C), 127.1 (1C), 126.6

(4C), 124.6 (4C), 123.3 (2C), 117.8 (1C), 96.9 (1C), 65.4 (1C), 65.1 (1C), 34.7 (2C), 31.7 (6C).

IR (ATR, ṽ) = 3121, 2959, 2927, 2914, 2900, 2864, 1601, 1507, 1486, 1459, 1431, 1421, 1392, 1360, 1322, 1295, 1278, 1266, 1204, 1187, 1166, 1142, 1115, 1069, 1028, 1017, 977, 966, 921, 899, 854, 834, 822, 749, 738, 728, 718, 708, 680, 667, 581, 565, 547, 527, 411 cm^{-1}.

FAB-MS (m/z, 3-NBA): 498/497 (52/100) [M]$^+$; HRMS–FAB (m/z): [M]$^+$ calcd for $C_{32}H_{35}O_2N_1S_1$, 497.2383, found 497.2382.

9-(4-(2-(3,4-Ethylenedioxy)thienyl)phenyl)-3,6-dimethoxycarbazole (29)

9-(4-Bromophenyl)-3,6-dimethoxycarbazole (9) (1.20 g, 3.14 mmol, 1.00 equiv), 2-pinacolboryl-3,4-ethylenedioxythiophene (16) (1.68 g, 6.28 mmol, 2.00 equiv), Pd(dppf)Cl$_2$ (115 mg, 157 µmol, 0.0500 equiv) and K$_2$CO$_3$ (2.60 g, 18.8 mmol, 6.00 equiv) were added to a 50 mL crimp cap vial and refilled with argon three times. 30 mL of degassed dimethyl sulfoxide was added, and the reaction mixture was heated to 90 °C. The reaction progress was monitored via TLC, and after 96 h, the product mixture was cooled to 22 °C. The mixture was poured into 100 mL water and extracted with ethyl acetate (3x 100 mL). The combined organic phases were washed with water (3x 200 mL) and brine (3x 200 mL), dried over anhydrous Na$_2$SO$_4$, and the solvents were removed under reduced pressure. The crude product was purified via column chromatography on silica gel (eluent cyclohexane/ethyl acetate 10:1) to yield 9-(4-(2-(3,4-ethylenedioxy)thienyl)phenyl)-3,6-dimethoxycarbazole (1.19 g, 2.68 mmol, 85% yield) as an off-white solid.

M.p.: 169 °C.

R$_f$ = 0.25 (cyclohexane/ethyl acetate 10:1).

^{1}H NMR (400 MHz, Dichloromethane-d2 [5.32 ppm], ppm) δ = 7.93 (d, J = 8.5 Hz, 2H), 7.57–7.53 (m, 4H), 7.37 (d, J = 8.9 Hz, 2H), 7.03 (dd, J = 2.4 Hz, J = 8.9 Hz, 2H), 6.36 (s, 1H), 4.32 (td, J = 3.4 Hz, J = 7.6 Hz, 4H), 3.93 (s, 6H).

^{13}C NMR (100 MHz, Dichloromethane-d2 [54.0 ppm], ppm) δ = 154.7 (2C), 143.1 (1C), 139.2 (1C), 136.8 (1C), 136.7 (2C), 132.4 (1C), 127.7 (2C), 127.1

(2C), 124.2 (2C), 116.9 (1C), 115.6 (2C), 111.3 (2C), 103.3 (2C), 98.4 (1C), 65.5 (1C), 65.1 (1C), 56.5 (2C).

IR (ATR, $\tilde{v}$) = 2932, 2830, 1601, 1578, 1519, 1487, 1470, 1456, 1431, 1411, 1361, 1329, 1310, 1292, 1264, 1248, 1205, 1169, 1152, 1123, 1103, 1068, 1027, 982, 956, 948, 922, 901, 841, 827, 805, 789, 751, 730, 708, 684, 673, 657, 636, 603, 588, 560, 543 cm^{-1}.

FAB-MS (m/z, 3-NBA): 444/443 (50/100) [M]$^+$, 154 (56), 137/136 (33/35); HRMS–FAB (m/z): [M]$^+$ calcd for $C_{26}H_{21}O_4N_1S_1$, 443.1186, found 443.1185.

4-(2-Selenyl)-N,N-bis(4-methoxyphenyl)aniline (30)

4-Bromo-N,N-bis(4-methoxyphenyl)aniline (1) (1.20 g, 3.12 mmol, 1.00 equiv), 2-pinacolborylselenophene (17) (963 mg, 3.75 mmol, 1.20 equiv), Pd(dppf)Cl$_2$ (114 mg, 156 µmol, 0.0500 equiv) and K$_2$CO$_3$ (2.59 g, 18.7 mmol, 6.00 equiv) were added to a 50 mL crimp cap vial and refilled with argon three times. 30 mL of degassed dimethyl sulfoxide was added, and the reaction mixture was heated to 90 °C. The reaction progress was monitored via TLC, and after 2 h, the product mixture was cooled to 22 °C. The mixture was poured into 100 mL water and extracted with ethyl acetate (3x 100 mL). The combined organic phases were washed with water (3x 200 mL) and brine (3x 200 mL), dried over anhydrous Na$_2$SO$_4$, and the solvents were removed under reduced pressure. The crude product was purified via column chromatography on silica gel (eluent cyclohexane/ethyl acetate 50:1 → 5:1 + 1% triethylamine) to yield 4-(2-selenyl)-N,N-bis(4-methoxyphenyl)aniline (1.14 g, 2.63 mmol, 84% yield) as an orange solid.

M.p.: 112 °C.

R$_f$ = 0.4 (cyclohexane/ethyl acetate 4:1).

^{1}H NMR (400 MHz, Dichloromethane-d2 [5.32 ppm], ppm) δ = 7.85 (d, J = 5.5 Hz, 1H), 7.37–7.33 (m, 3H), 7.27 (dd, J = 4.0 Hz, J = 5.4 Hz, 1H), 7.06 (d, J = 8.9 Hz, 4H), 6.88–6.84 (m, 6H), 3.79 (s, 6H).

^{13}C NMR (100 MHz, Dichloromethane-d2 [54.0 ppm], ppm) δ = 156.8 (2C), 151.4 (1C), 149.0 (1C), 141.1 (2C), 131.1 (1C), 129.0 (1C), 128.9 (1C), 127.4 (2C), 127.3 (4C), 124.2 (1C), 120.7 (2C), 115.2 (4C), 56.0 (2C).

IR (ATR, ṽ) = 3105, 3031, 3002, 2945, 2922, 2901, 2853, 2830, 1864, 1731, 1598, 1497, 1452, 1439, 1425, 1330, 1296, 1279, 1268, 1241, 1183, 1160, 1130, 1101, 1084, 1034, 1007, 955, 932, 909, 826, 813, 789, 734, 715, 686, 649, 633, 599, 569, 548, 523 cm^{-1}.

FAB-MS (m/z, 3-NBA): 436/435/433 (40/100/53) [M]$^+$; HRMS–FAB (m/z): [M]$^+$ calcd for C$_{24}$H$_{21}$N$_1$Se$_1$, 435.0732, found 435.0731.

4,16-Di(4-(2-thienyl)-N,N-bis(4-methoxy-phenyl)aniline)[2.2]paracyclophane (TM_31)

Via inside-out approach:

4,16-Dithienyl-[2.2]paracyclophane (11) (100 mg, 268 µmol, 1.00 equiv), 4-bromo-N,N-bis(4-methoxyphenyl)aniline (1) (413 mg, 1.07 mmol, 4.00 equiv), Pd(OAc)$_2$ (3.01 mg, 13.4 µmol, 0.0500 equiv), PPh$_3$ (7.04 mg, 26.8 µmol, 0.100 equiv), PivOH (8.22 mg, 80.5 µmol, 0.300 equiv) and K$_2$CO$_3$ (55.6 mg, 403 µmol, 1.50 equiv) were suspended in 3 mL of N,N-dimethylformamide in an argon atmosphere and the reaction mixture was heated to 100 °C. The reaction progress was monitored via TLC. After 13 days, the product mixture was cooled to 22 °C, poured into 50 mL water, and extracted with dichloromethane (3x 50 mL). The combined organic phases were washed with water (3x 150 mL) and brine (3x 150 mL) and dried over anhydrous Na$_2$SO$_4$. After removing the solvent under reduced pressure, the crude product was purified via column chromatography on silica gel (eluent cyclohexane/ethyl acetate 100:1 → 10:1) to yield 4,16-di(4-(2-thienyl)-N,N-bis(4-methoxy-phenyl)aniline)[2.2]paracyclophane (100 mg, 102 µmol, 38% yield) as a yellow solid.

Via outside-in approach:

4,16-Dibromo[2.2]paracyclophane (200 mg, 546 µmol, 1.00 equiv), 4-(2-thienyl)-*N,N*-bis(4-methoxyphenyl)aniline (12) (466 mg, 1.20 mmol, 2.20 equiv), Pd(OAc)$_2$ (6.13 mg, 27.3 µmol, 0.0500 equiv), PPh$_3$ (14.3 mg, 54.6 µmol, 0.100 equiv), K$_2$CO$_3$ (113 mg, 819 µmol, 1.50 equiv) and pivalic acid (16.7 mg, 164 µmol, 0.300 equiv) were added into a crimp cap vial, evacuated and backfilled with argon for three times. Then 10 mL of dry *N,N*-dimethylformamide was added, and the reaction mixture was heated to 100 °C. The reaction progress was monitored *via* TLC. After 17 h, the product mixture was cooled to 22 °C, poured into 50 mL water, and extracted with dichloromethane (3x 50 mL). The combined organic phases were washed with water (3x 150 mL) and brine (3x 150 mL) and dried over anhydrous Na$_2$SO$_4$. After removing the solvent under reduced pressure, the crude product was purified *via* column chromatography on silica gel (eluent cyclohexane/ethyl acetate 100:1 → 10:1) to yield 4,16-di(4-(2-thienyl)-*N,N*-bis(4-methoxyphenyl)aniline)[2.2]paracyclophane (330 mg, 337 µmol, 62% yield) as a yellow solid.

M.p.: >200 °C.

R$_f$ = 0.18 (cyclohexane/ethyl acetate 10:1).

^{1}H NMR (400 MHz, Dichloromethane-d2 [5.32 ppm], ppm) δ = 7.50 (d, *J* = 8.5 Hz, 4H), 7.25 (d, *J* = 3.4 Hz, 2H), 7.09 (d, *J* = 8.8 Hz, 10H), 6.94 (d, *J* = 8.4 Hz, 4H), 6.87 (d, *J* = 8.6 Hz, 8H), 6.76 (d, *J* = 7.5 Hz, 2H), 6.68 (s, 2H), 6.61 (d, *J* = 7.8 Hz, 2H), 3.80 (s, 14H), 3.00–2.95 (m, 2H), 2.92–2.91 (m, 4H).

^{13}C NMR (100 MHz, Dichloromethane-d2 [54.0 ppm], ppm) δ = 156.8 (4C), 148.9 (2C), 144.9 (2C), 142.8 (2C), 141.1 (4C), 140.8 (2C), 137.7(2C), 135.6 (2C), 135.5 (2C), 133.6 (2C), 130.1 (2C), 127.6 (2C), 127.3 (8C), 126.9 (2C), 126.7 (4C), 122.8 (2C), 120.9 (4C), 115.3 (8C), 56.0 (4C), 34.8 (2C), 34.7 (2C).

IR (ATR, ṽ) = 3034, 2996, 2929, 2833, 1601, 1500, 1483, 1459, 1438, 1402, 1317, 1282, 1265, 1235, 1190, 1177, 1164, 1103, 1031, 952, 824, 800, 779, 728, 713, 701, 577, 523 cm^{-1}.

ESI-MS (*m/z*): 979/978 (43/61) [M]$^+$, 772 (32), 490/489 (70/100); HRMS–ESI (*m/z*): [M]$^+$ calcd for C$_{64}$H$_{54}$N$_2$O$_2$S$_2$, 978.3525; found, 978.3518.

4,16-Di(4-(2-thienyl)-*N,N*-bis(phenyl)aniline)[2.2]paracyclophane (TM_32)

4,16-Dibromo[2.2]paracyclophane (100 mg, 273 μmol, 1.00 equiv), 4-(2-thienyl)-*N,N*-bis(phenyl)aniline (13) (358 mg, 1.09 mmol, 4.00 equiv), Pd(OAc)$_2$ (3.07 mg, 13.7 μmol, 0.0500 equiv), PPh$_3$ (7.16 mg, 27.3 μmol, 0.100 equiv), K$_2$CO$_3$ (56.6 mg, 410 μmol, 1.50 equiv) and pivalic acid (8.37 mg, 81.9 μmol, 0.300 equiv) were added into a crimp cap vial, evacuated and backfilled with argon for three times. Then 5 mL of dry *N,N*-dimethylformamide was added, and the reaction mixture was heated to 100 °C. The reaction progress was monitored *via* TLC, and after 44 h, the mixture was cooled to 22 °C. The mixture was poured into water (50 mL) and extracted with dichloromethane (3x 50 mL). The combined organic phases were washed with water (3x 150 mL) and brine (3x 150 mL), dried over anhydrous Na$_2$SO$_4$, and the solvent was removed under reduced pressure. The crude product was purified *via* column chromatography on silica gel (eluent cyclohexane/ethyl acetate 100:1 → 80:1) and afterward washed with *n*-heptane to yield 4,16-di(4-(2-thienyl)-*N,N*-bis(phenyl)aniline)[2.2]paracyclophane (140 mg, 163 μmol, 60% yield) as a yellow solid.

M.p.: >200 °C.

R_f = 0.45 (cyclohexane/ethyl acetate 40:1).

^{1}H NMR (400 MHz, Chloroform-d [7.27 ppm], ppm) δ = 7.60–7.56 (m, 4H), 7.32–7.31 (m, 2H), 7.30–7.28 (m, 8H), 7.17–7.12 (m, 12H), 7.09–7.04 (m, 6H), 6.80 (dd, *J* = 1.8 Hz, *J* = 7.8 Hz, 2H), 6.69 (d, *J* = 1.8 Hz, 2H), 6.63 (d, *J* = 7.8 Hz, 2H), 3.87–3.80 (m, 2H), 3.04–2.98 (m, 2H), 2.97–2.92 (m, 4H).

^{13}C NMR (100 MHz, Chloroform-d [77.0 ppm], ppm) δ = 147.5 (4C), 147.2 (2C), 143.9 (2C), 142.8 (2C), 140.1 (2C), 137.1 (2C), 135.0 (2C), 135.0 (2C), 133.2 (2C), 129.6 (2C), 129.3 (8C), 128.6 (2C), 127.0 (2C), 126.4 (4C), 124.4 (8C), 123.9 (4C), 123.1 (4C), 122.8 (2C), 34.4 (2C), 34.2 (2C).

IR (ATR, ṽ) = 3058, 3029, 2968, 2925, 2885, 2847, 1587, 1540, 1507, 1485, 1460, 1451, 1401, 1312, 1289, 1268, 1207, 1188, 1177, 1153, 1109, 1072, 1057, 1026, 994, 948, 897, 823, 800, 745, 725, 693, 666, 654, 639, 615, 595, 567 cm^{-1}.

ESI-MS (m/z): 860/859/858 (18/64/100) $[M]^+$, 698/697 (11/24), 610/609 (12/26), 534/533 (17/43), 489/488 (22/59), 430/429 (56/85), 327 (23); HRMS–ESI (m/z): $[M]^+$ calcd for $C_{60}H_{46}N_2S_2$, 858.3102; found, 858.3094.

4,16-Di(4-(2-(3,4-ethylenedioxy)thienyl)-*N,N*-bis(4-methoxy-phenyl)aniline)[2.2]paracyclophane (TM_33)

4,16-Dibromo[2.2]paracyclophane (150 mg, 410 µmol, 1.00 equiv), 4-(2-(3,4-ethylenedioxy)thienyl)-*N,N*-bis(4-methoxyphenyl)aniline (14) (402 mg, 901 µmol, 2.20 equiv), Pd(OAc)$_2$ (9.20 mg, 41.0 µmol, 0.100 equiv), PPh$_3$ (21.5 mg, 81.9 µmol, 0.200 equiv), K$_2$CO$_3$ (170 mg, 1.23 mmol, 3.00 equiv) and pivalic acid (25.1 mg, 246 µmol, 0.600 equiv) were added into a crimp cap vial, evacuated and backfilled with argon for three times. Then 10 mL of dry *N,N*-dimethylformamide was added, and the reaction mixture was heated to 120 °C. The reaction progress was monitored *via* TLC. After 18 h, the product mixture was cooled to 23 °C, poured into 50 mL water, and extracted with dichloromethane (3x 100 mL). The combined organic phases were washed with water (3x 150 mL) and brine (3x 150 mL), dried over anhydrous Na$_2$SO$_4$, and the solvent was removed under reduced pressure. The crude product was purified *via* column chromatography on silica gel (eluent dichloromethane + 1% triethylamine) and by wasing with water to yield 4,16-di(4-(2-(3,4-ethylenedioxy)thienyl)-*N,N*-bis(4-methoxyphenyl)aniline)[2.2]paracyclophane (168 mg, 153 µmol, 37% yield) as a yellow solid.

M.p.: >200 °C.

R_f = 0.22 (cyclohexane/ethyl acetate 4:1).

^{1}H NMR (400 MHz, Chloroform-d [7.27 ppm], ppm) δ = 7.65 (d, J = 8.8 Hz, 4H), 7.09 (d, J = 8.9 Hz, 8H), 6.99 (d, J = 8.8 Hz, 4H), 6.85 (d, J = 8.9 Hz, 8H), 6.80 (dd, J = 1.3 Hz, J = 7.8 Hz, 2H), 6.64 (d, J = 1.3 Hz, 2H), 6.59 (d, J = 7.9

Hz, 2H), 4.40–4.21 (m, 8H), 3.82 (s, 12H), 3.53–3.46 (m, 2H), 3.06–2.99 (m, 2H), 2.95–2.90 (m, 4H).

^{13}C NMR (100 MHz, Chloroform-d [77.0 ppm], ppm) δ = 155.8 (4C), 147.3 (2C), 147.2 (2C), 140.9 (4C), 140.0 (2C), 138.5 (2C), 137.2 (2C), 136.9 (2C), 134.8 (2C), 133.1 (2C), 130.1 (2C), 126.7 (4C), 126.4 (8C), 125.5 (2C), 121.0 (4C), 116.7 (2C), 115.2 (2C), 114.7 (8C), 64.7 (2C), 64.5 (2C), 55.5 (4C), 34.7 (2C), 34.5 (2C).

IR (ATR, $\tilde{v}$) = 2944, 2924, 2832, 1606, 1589, 1521, 1503, 1493, 1460, 1452, 1438, 1402, 1360, 1317, 1290, 1261, 1238, 1191, 1181, 1164, 1132, 1105, 1085, 1033, 990, 976, 958, 938, 915, 895, 864, 840, 823, 810, 782, 730, 707, 679, 663, 645, 637, 598, 592, 574, 551, 528 cm^{-1}.

ESI-MS (m/z): 1095/1094 (30/39) [M]$^+$, 548/547 (71/100) [M]$^{2+}$; HRMS–ESI (m/z): [M]$^+$ calcd for C$_{68}$H$_{58}$N$_2$O$_8$S$_2$, 1094.3635; found, 1094.3631.

4,16-Di(4-(2-thienyl)-N,N-bis(3,4,5-methoxy-phenyl)aniline)[2.2]paracyclophane (TM_34)

4,16-Di-bromo[2.2]paracyclophane (120 mg, 328 µmol, 1.00 equiv), 4-(2-thienyl)-N,N-bis(3,4,5-trimethoxy-phenyl)aniline (21) (366 mg, 721 µmol, 2.20 equiv), Pd(OAc)$_2$ (7.36 mg, 32.8 µmol, 0.100 equiv), PPh$_3$ (17.2 mg, 65.6 µmol, 0.200 equiv), K$_2$CO$_3$ (136 mg, 983 µmol, 3.00 equiv) and pivalic acid (20.1 mg, 197 µmol, 0.600 equiv) were added into a crimp cap vial, evacuated and backfilled with argon for three times. Then 6 mL of dry N,N-dimethylformamide was added, and the reaction mixture was heated to 120 °C. The reaction progress was monitored via TLC. After 5 days, the product mixture was cooled to 23 °C, poured into 50 mL water, and extracted with dichloromethane (3x 100 mL). The combined organic phases were washed with water (3x 150 mL) and brine (3x 150 mL), dried over anhydrous Na$_2$SO$_4$, and the solvent was removed under reduced pressure. The crude product was purified via column chromatography on silica gel (eluent cyclohexane/dichloromethane

1:1 + 1% triethylamine). After removing the solvent under reduced pressure, the crude product was washed with cyclohexane twice to yield 4,16-di(4-(2-thienyl)-*N,N*-bis(3,4,5-methoxyphenyl)aniline)[2.2]paracyclophane (190 mg, 156 μmol, 48% yield) as a yellow solid.

M.p.: >200 °C.

R_f = 0.05 (cyclohexane/ethyl acetate 4:1).

^{1}H NMR (400 MHz, Dichloromethane-d2 [5.32 ppm], ppm) δ = 7.57 (d, *J* = 8.5 Hz, 4H), 7.30 (d, *J* = 3.5 Hz, 2H), 7.11–7.07 (m, 6H), 6.77 (d, *J* = 7.8 Hz, 2H), 6.69 (s, 2H), 6.63 (d, *J* = 7.8 Hz, 2H), 6.38 (s, 8H), 3.78 (s, 14H), 3.72 (s, 24H), 3.04–2.98 (m, 2H), 2.94–2.89 (m, 4H).

^{13}C NMR (100 MHz, Dichloromethane-d2 [54.0 ppm], ppm) δ = 154.4 (8C), 147.9 (2C), 144.6 (2C), 143.7 (4C), 143.2 (2C), 140.8 (2C), 137.8 (2C), 135.6 (2C), 135.5 (2C), 135.2 (4C), 133.6 (2C), 130.1 (2C), 128.3 (2C), 127.7 (2C), 126.6 (4C), 123.2 (4C), 123.2 (2C), 103.3 (8C), 61.1 (4C), 56.6 (8C), 34.8 (2C), 34.7 (2C).

IR (ATR, ṽ) = 2990, 2931, 2827, 1584, 1496, 1446, 1425, 1408, 1347, 1265, 1227, 1205, 1181, 1122, 1101, 1050, 1003, 953, 924, 899, 861, 819, 802, 769, 752, 730, 711, 694, 645, 605, 569, 523 cm^{-1}.

ESI-MS (*m/z*): 1219/1218 (2/3) [M]$^+$, 610/609 (10/12), 508/507 (2/6), 337/335 (2/3), 152/151/150 (31/7/100), 118/116 (2/6), 102 (14); HRMS–ESI (*m/z*): [M]$^+$ calcd for $C_{72}H_{70}N_2O_{12}S_2$, 1218.4370; found, 1218.4351.

4,16-Di(4-(2-thienyl)-*N,N*-bis(4-*tert*-butyl-

phenyl)aniline)[2.2]paracyclophane (TM_35)

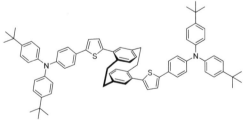

4,16-Di-bromo[2.2]paracyclophane (200 mg, 546 μmol, 1.00 equiv), 4-(2-thienyl)-*N,N*-bis(4-*tert*-butylphenyl)aniline (22) (528 mg, 1.20 mmol, 2.20 equiv), Pd(OAc)$_2$ (6.13 mg, 27.3 μmol, 0.0500 equiv), PPh$_3$ (14.3 mg, 54.6 μmol, 0.100 equiv), K$_2$CO$_3$ (113

mg, 819 μmol, 1.50 equiv) and pivalic acid (16.7 mg, 164 μmol, 0.300 equiv) were added into a crimp cap vial, evacuated and backfilled with argon for three times. Then 10 mL of dry *N,N*-dimethylformamide was added, and the reaction mixture was heated to 100 °C. The reaction progress was monitored *via* TLC. After 16 h, the product mixture was cooled to 22 °C and poured into 50 mL water, extracted with dichloromethane (3x 50 mL). The combined organic phases were washed with water (3x 150 mL) and brine (3x 150 mL) and dried over anhydrous Na_2SO_4. After removing the solvent under reduced pressure, the crude product was purified *via* column chromatography on silica gel (eluent cyclohexane/ethyl acetate 100:1) and precipitation from a cyclohexane/acetonitrile mixture (1:1) to yield 4,16-di(4-(2-thienyl)-*N,N*-bis(4-*tert*-butylphenyl)aniline)[2.2]paracyclophane (280 mg, 258 μmol, 47% yield) as a yellow solid.

M.p.: >200 °C.

R_f = 0.4 (cyclohexane/ethyl acetate 10:1).

^1H NMR (400 MHz, Chloroform-d [7.27 ppm], ppm) δ = 7.57–7.54 (m, 4H), 7.31–7.27 (m, 10H), 7.12–7.07 (m, 14H), 6.81–6.79 (m, 2H), 6.69 (d, *J* = 1.3 Hz, 2H), 6.62 (d, *J* = 7.8 Hz, 2H), 3.86–3.80 (m, 2H), 3.03–2.97 (m, 2H), 2.96–2.92 (m, 4H), 1.34 (s, 36H).

^{13}C NMR (100 MHz, Chloroform-d [77.0 ppm], ppm) δ = 147.6 (2C), 145.9 (4C), 144.8 (4C), 144.2 (2C), 142.5 (2C), 140.1 (2C), 137.1 (2C), 135.0 (2C), 133.2 (2C), 129.6 (2C), 127.8 (2C), 127.7 (2C), 127.0 (2C), 126.2 (4C), 126.1 (8C), 124.1 (8C), 123.0 (4C), 122.5 (2C), 34.4 (2C), 34.3 (2C), 34.2 (4C), 31.4 (12C).

IR (ATR, ṽ) = 3031, 2959, 2927, 2901, 2864, 1598, 1538, 1504, 1479, 1460, 1395, 1363, 1320, 1298, 1288, 1265, 1203, 1181, 1108, 1062, 1016, 955, 899, 826, 799, 752, 737, 722, 652, 581, 574, 564, 552, 543 cm^{-1}.

ESI-MS (*m/z*): 1085/1084/1083 (13/37/45) [M]$^+$, 645 (21), 542/541 (68/78), 440/439 (31/100), 316 (20), 288 (55); HRMS–ESI (*m/z*): [M]$^+$ calcd for $C_{76}H_{78}N_2S_2$, 1082.5606; found, 1082.5585.

4,16-Di(9-(4-(2-thienyl)phenyl)carbazole)[2.2]paracyclophane (TM_36)

4,16-Dibromo[2.2]paracyclophane (200 mg, 546 µmol, 1.00 equiv), 9-(4-(2-thienyl)phenyl)carbazole (24) (391 mg, 1.20 mmol, 2.20 equiv), Pd(OAc)₂ (6.13 mg, 27.3 µmol, 0.0500 equiv), PPh₃ (14.3 mg, 54.6 µmol, 0.100 equiv), K₂CO₃ (113 mg, 819 µmol, 1.50 equiv) and pivalic acid (16.7 mg, 164 µmol, 0.300 equiv) were added into a crimp cap vial, evacuated and backfilled with argon for three times. Then 10 mL of dry N,N-dimethylformamide was added, and the reaction mixture was heated to 100 °C. The reaction progress was monitored *via* TLC. After 16 h, the product mixture was cooled to 22 °C, poured into 50 mL water, and extracted with dichloromethane (3x 50 mL). The combined organic phases were washed with water (3x 150 mL) and brine (3x 150 mL) and dried over anhydrous Na₂SO₄. After removing the solvent under reduced pressure, the crude product was purified *via* column chromatography on silica gel (eluent cyclohexane/ethyl acetate 100:1) and by precipitation from cyclohexane/dichloromethane (1:1) to yield 4,16-di(9-(4-(2-thienyl)phenyl)carbazole)[2.2]paracyclophane (183 mg, 214 µmol, 39% yield) as a yellow solid.

M.p.: >200 °C.

R_f = 0.2 (cyclohexane/ethyl acetate 10:1).

¹H NMR (400 MHz, Chloroform-d [7.27 ppm], ppm) δ = 8.20–8.18 (m, 4H), 7.97–7.94 (m, 4H), 7.67–7.64 (m, 4H), 7.52–7.44 (m, 10H), 7.35–7.31 (m, 4H), 7.19 (d, J = 3.5 Hz, 2H), 6.88 (dd, J = 1.8 Hz, J = 7.8 Hz, 2H), 6.77 (d, J = 1.8 Hz, 2H), 6.70 (d, J = 7.8 Hz, 2H), 3.92–3.86 (m, 2H), 3.11–3.04 (m, 2H), 3.02–2.99 (m, 4H).

¹³C NMR (100 MHz, Chloroform-d [77.0 ppm], ppm) δ = 144.2 (2C), 143.0 (2C), 140.8 (4C), 140.2 (2C), 137.3 (2C), 136.8 (2C), 135.1 (2C), 134.8 (2C), 133.5 (2C), 133.3 (2C), 129.9 (2C), 127.5 (4C), 127.3 (2C), 126.9 (4C), 126.0 (4C), 124.2 (2C), 123.5 (4C), 120.4 (4C), 120.1 (4C), 109.8 (4C), 34.5 (2C), 34.2 (2C).

IR (ATR, $\tilde{v}$) = 3043, 2919, 2847, 1623, 1598, 1540, 1511, 1489, 1479, 1449, 1415, 1402, 1363, 1346, 1333, 1316, 1300, 1273, 1230, 1204, 1187, 1170, 1149, 1118, 1109, 1060, 1016, 999, 914, 901, 834, 820, 802, 745, 721, 686, 671, 653, 643, 625, 565, 531 cm^{-1}.

ESI–MS (*m/z*): 855/854 (20/31) [M]$^+$, 684/683/682/681 (13/29/45/100), 564 (12), 448 (10), 428/427 (15/24), 316 (28), 289/288 (12/66), 282 (10), 123 (12); HRMS–ESI (*m/z*): [M]$^+$ calcd for $C_{60}H_{42}N_2S_2$, 854.2789; found, 854.2771.

4,16-Di(9-(4-(2-thienyl)phenyl)-3,6-dimethoxycarb-azole)[2.2]paracyclophane (TM_37)

4,16-Di-bromo[2.2]paracyclophane (200 mg, 546 µmol, 1.00 equiv), 9-(4-(2-thienyl)phenyl)-3,6-di-methoxycarbazole (25) (463 mg, 1.20 mmol, 2.20 equiv), Pd(OAc)$_2$ (6.13 mg, 27.3 µmol, 0.0500 equiv), PPh$_3$ (14.3 mg, 54.6 µmol, 0.100 equiv), K$_2$CO$_3$ (113 mg, 819 µmol, 1.50 equiv) and pivalic acid (16.7 mg, 164 µmol, 0.300 equiv) were added into a crimp cap vial, evacuated and backfilled with argon for three times. Then 10 mL of dry *N,N*-dimethylformamide was added, and the reaction mixture was heated to 100 °C. The reaction progress was monitored *via* TLC and stirred for 64 h before cooling to 22 °C. The cooled mixture was poured into 50 mL water and extracted with dichloromethane (3x 50 mL). The combined organic phases were washed with water (3x 150 mL) and brine (3x 150 mL) and dried over anhydrous Na$_2$SO$_4$. The solvent was removed under reduced pressure, and the crude product was purified *via* column chromatography on silica gel (eluent cyclo-hexane/ethyl acetate 100:1 → 8:1) and afterward precipitated from a cyclo-hexane/dichloromethane (5:1) mixture to yield 4,16-di(9-(4-(2-thienyl)phenyl)-3,6-dimethoxycarbazole)[2.2]paracyclophane (170 mg, 174 µmol, 32% yield) as a yellow solid.

M.p.: >200 °C.

R$_f$ = 0.33 (cyclohexane/ethyl acetate 10:1).

^{1}H NMR (400 MHz, Chloroform-d [7.27 ppm], ppm) δ = 7.93–7.90 (m, 4H), 7.63–7.61 (m, 4H), 7.59 (d, J = 2.5 Hz, 4H), 7.46 (d, J = 3.8 Hz, 2H), 7.43–7.40 (m, 4H), 7.18 (d, J = 3.8 Hz, 2H), 7.10–7.07 (m, 4H), 6.87 (dd, J = 1.8 Hz, J = 7.8 Hz, 2H), 6.76 (d, J = 1.8 Hz, 2H), 6.69 (d, J = 7.8 Hz, 2H), 3.98 (s, 12H), 3.91–3.85 (m, 2H), 3.10–3.03 (m, 2H), 3.01–2.98 (m, 4H).

^{13}C NMR (100 MHz, Chloroform-d [77.0 ppm], ppm) δ = 154.1 (4C), 144.0 (2C), 143.1 (2C), 140.2 (2C), 137.3 (2C), 137.2 (2C), 136.1 (4C), 135.1 (2C), 134.8 (2C), 133.3 (2C), 133.0 (2C), 129.8 (2C), 127.3 (2C), 127.0 (4C), 126.9 (4C), 124.0 (2C), 123.7 (4C), 115.2 (4C), 110.7 (4C), 103.0 (4C), 56.1 (4C), 34.4 (2C), 34.2 (2C).

IR (ATR, $\tilde{v}$) = 2919, 2897, 2891, 2826, 1605, 1582, 1540, 1510, 1487, 1463, 1431, 1402, 1367, 1327, 1305, 1285, 1264, 1203, 1174, 1154, 1106, 1064, 1033, 946, 908, 897, 836, 824, 798, 790, 751, 725, 659, 636, 605, 584, 567, 535 cm^{-1}.

ESI-MS (m/z): 975/974 (14/21) [M]$^+$, 684/683/682/681 (14/32/47/100), 488/487 (67/96), 316 (34), 289/288 (15/85); HRMS–ESI (m/z): [M]$^+$ calcd for C$_{64}$H$_{75}$N$_2$O$_4$S$_2$, 974.3212; found, 974.3188.

4,16-Di(9-(4-(2-thienyl)phenyl)-3,6-di*tert*-butylcarb-azole)[2.2]paracyclophane (TM_38)

4,16-Di-bromo[2.2]paracyclophane (200 mg, 546 µmol, 1.00 equiv), 9-(4-(2-thienyl)phenyl)-3,6-di*tert*-butylcarbazole (26) (526 mg, 1.20 mmol, 2.20 equiv), Pd(OAc)$_2$ (6.13 mg, 27.3 µmol, 0.0500 equiv), PPh$_3$ (14.3 mg, 54.6 µmol, 0.100 equiv), K$_2$CO$_3$ (113 mg, 819 µmol, 1.50 equiv) and pivalic acid (16.7 mg, 164 µmol, 0.300 equiv) were added into a crimp cap vial, evacuated and backfilled with argon for three times. Then 10 mL of dry *N,N*-dimethylformamide was added, and the reaction mixture was heated to 100 °C. The reaction progress was monitored *via* TLC. After 16 h, the product mixture was cooled to 22 °C, poured into 50 mL water, and extracted with dichloromethane (3x 50 mL). The combined organic phases were washed with water (3x 150 mL) and brine (3x 150 mL) and dried over anhydrous Na$_2$SO$_4$. After removing the solvent under reduced pressure, the crude product was purified *via* column chromatography on silica gel (eluent cyclohexane/ethyl acetate 100:1) and by precipitation from heptane/dichloromethane (10:1) to yield 4,16-di(9-(4-(2-thienyl)phenyl)-3,6-di*tert*-butylcarbazole)[2.2]paracyclophane (140 mg, 130 µmol, 24% yield) as a yellow solid.

M.p.: >200 °C.

R$_f$ = 0.2 (cyclohexane/ethyl acetate 10:1).

^{1}H NMR (400 MHz, Chloroform-d [7.27 ppm], ppm) δ = 8.18 (d, J = 1.5 Hz, 4H), 7.93 (d, J = 8.3 Hz, 4H), 7.64 (d, J = 8.6 Hz, 4H), 7.53–7.43 (m, 10H), 7.18 (d, J = 3.8 Hz, 2H), 6.88 (dd, J = 1.8 Hz, J = 7.8 Hz, 2H), 6.77 (d, J = 1.5 Hz, 2H), 6.70 (d, J = 8.1 Hz, 2H), 3.92–3.85 (m, 2H), 3.11–3.04 (m, 2H), 3.02–2.98 (m, 4H), 1.50 (s, 36H).

^{13}C NMR (100 MHz, Chloroform-d [77.0 ppm], ppm) δ = 144.0 (2C), 143.2 (2C), 143.0 (4C), 140.2 (2C), 139.1 (4C), 137.3 (2C), 137.2 (2C), 135.1 (2C), 134.8 (2C), 133.3 (2C), 133.0 (2C), 129.8 (2C), 127.3 (2C), 127.1 (4C), 126.8 (4C), 124.0 (2C), 123.7 (4C), 123.4 (4C), 116.3 (4C), 109.2 (4C), 34.8 (4C), 34.4 (2C), 34.2 (2C), 32.0 (12C).

IR (ATR, ṽ) = 2958, 2902, 2863, 1596, 1541, 1511, 1486, 1472, 1453, 1363, 1323, 1293, 1259, 1231, 1201, 1112, 1106, 1030, 955, 878, 839, 805, 739, 730, 697, 652, 611, 596 cm^{-1}.

ESI-MS (*m/z*): 1080/1079 (8/10) [M]$^+$, 684/683/682/681 (13/30/45/100), 644/643 (8/17), 565/564 (9/25), 540/539 (25/32), 513 (16), 484 (8), 438/437 (24/87), 316 (36), 289/288 (16/90), 283/282 (10/47), 280 (10), 123 (8); HRMS–ESI (*m/z*): [M]$^+$ calcd for C$_{76}$H$_{74}$N$_2$S$_2$, 1078.5293; found, 1078.5275.

5.2.3 Bis(pseudo-*meta*)-*para*-substituted [2.2]Paracyclophanes

4,7,13,16-Tetra(4-(2-thienyl)-*N*,*N*-bis(4-methoxy-phenyl)aniline)[2.2]paracyclophane (TM_39)

4-(2-Thienyl)-*N*,*N*-bis(4-methoxy-phenyl)-aniline (12) (533 mg, 1.37 mmol, 4.80 equiv) was added to a crimp cap vial and flushed with argon three times. 5 mL of dry tetrahydrofuran was added, and the mixture was cooled to -78 °C. *n*-Butyllithium solution (96.8 mg, 605 µL, 1.51 mmol, 2.50M, 5.28 equiv) was added dropwise, and the reaction mixture was warmed to 22 °C. After stirring for 30 min, the reaction mixture was cooled to 0 °C, and zinc chloride (206 mg, 1.51 mL, 1.51 mmol, 1.00M, 5.28 equiv) in tetrahydrofuran was added. The reaction was stirred for 15 min at 0 °C and then warmed to 22 °C. 4,7,13,16-Tetrabromo[2.2]paracyclophane (150 mg, 286 µmol, 1.00 equiv) and PEPPSI™-iPr (15.6 mg, 22.9 µmol, 0.0800 equiv) was added to a crimp cap vial and flushed with argon three times. 5 mL of dry tetrahydrofuran and zinc chloride (156 mg, 1.15 mL, 1.15 mmol, 1.00M, 4.00 equiv) in tetrahydrofuran were added. After dissolving, the NEGISHI reactant was added *via* a syringe, and the reaction mixture was heated to 70 °C. The reaction progress was monitored *via* TLC. After 18 h, the reaction was cooled to 22 °C, filtered through a pad of celite, and the solvent was removed under reduced pressure. The crude mixture was purified *via* column chromatography on silica gel (eluent cyclohexane/dichloromethane/ethyl acetate 7:1:0 → 0:4:1) to yield 4,7,13,16-tetra(4-(2-thienyl)-*N*,*N*-bis(4-methoxy-phenyl)aniline)[2.2]paracyclophane (320 mg, 183 µmol, 64% yield) as a yellow solid.

M.p.: >200 °C.

R_f = 0.1 (cyclohexane/ethyl acetate 4:1).

¹H NMR (400 MHz, Dichloromethane-d2 [5.32 ppm], ppm) δ = 7.47–7.43 (m, 8H), 7.24 (d, *J* = 3.8 Hz, 4H), 7.14 (d, *J* = 3.8 Hz, 4H), 7.07–7.04 (m, 16H), 7.00 (s, 4H), 6.89–6.87 (m, 8H), 6.85–6.83 (m, 16H), 3.78–3.77 (m, 28H), 3.02–2.94 (m, 4H).

¹³C NMR (100 MHz, Dichloromethane-d2 [54.0 ppm], ppm) δ = 156.7 (8C), 148.9 (4C), 144.9 (4C), 141.6 (4C), 141.2 (4C), 141.1 (8C), 137.2 (4C), 134.4 (4C), 132.9 (4C), 127.2 (16C), 126.9 (4C), 126.6 (8C), 122.7 (4C), 121.2 (8C), 115.2 (16C), 56.0 (8C), 34.1 (4C).

IR (ATR, ṽ) = 2944, 2929, 2902, 2830, 1599, 1499, 1482, 1456, 1438, 1316, 1279, 1234, 1190, 1176, 1163, 1103, 1031, 956, 911, 823, 795, 720, 701, 683, 598, 574 cm⁻¹.

ESI-MS (*m/z*): 1751/1750/1749 (1/1/1) [M]⁺, 877/876/875/874 (1/17/49/40) [M]²⁺, 584/583 (19/26) [M]³⁺, 460/459/458/457 (1/4/29/100); HRMS–ESI (*m/z*): [M]⁺ calcd for $C_{112}H_{92}N_4O_8S_4$, 1748.5798; found, 1748.5778.

4,7,13,16-Tetra(4-(2-thienyl)-*N,N*-bis(phenyl)aniline)[2.2]paracyclophane (TM_40)

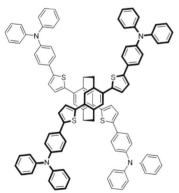

4-(2-Thienyl)-*N,N*-bis(phenyl)aniline (13) (450 mg, 1.37 mmol, 4.80 equiv) was added to a crimp cap vial and flushed with argon three times. 5 mL of dry tetrahydrofuran was added, and the mixture was cooled to -78 °C. *n*-Butyllithium solution (96.8 mg, 605 μL, 1.51 mmol, 2.50M, 5.28 equiv) was added dropwise, and the reaction mixture was warmed to 22 °C. After stirring for 30 min, the reaction mixture was cooled to 0 °C, and zinc chloride (206 mg, 1.51 mL, 1.51 mmol, 1.00M, 5.28 equiv) in tetrahydrofuran was added. The reaction was stirred for 15 min at 0 °C and then warmed to 22 °C. 4,7,13,16-Tetrabromo[2.2]paracyclophane (150 mg, 286 μmol, 1.00 equiv) and PEPPSI™-iPr (15.6 mg, 22.9 μmol, 0.0800 equiv) was added to a crimp cap vial and flushed with argon three times. 5 mL of dry tetrahydrofuran and zinc chloride (156 mg,

173

1.15 mL, 1.15 mmol, 1.00M, 4.00 equiv) in tetrahydrofuran were added. After dissolving, the NEGISHI reactant was added *via* a syringe, and the reaction mixture was heated to 70 °C. The reaction progress was monitored *via* TLC. After 19 h, the reaction was cooled to 22 °C, filtered through a pad of celite, and the solvent was removed under reduced pressure. The crude mixture was purified *via* column chromatography on silica gel (eluent cyclohexane/dichloromethane 20:1 → 2:1) to yield 4,7,13,16-tetra(4-(2-thienyl)-*N,N*-bis(phenyl)aniline)[2.2]paracyclophane (228 mg, 151 µmol, 53% yield) as a yellow solid.

M.p.: >200 °C.

R_f = 0.25 (cyclohexane/dichloromethane 10:1).

^{1}H NMR (400 MHz, Chloroform-d [7.27 ppm], ppm) δ = 7.52 (d, *J* = 8.4 Hz, 8H), 7.29 (s, 10H), 7.25 (s, 10H), 7.13 (d, *J* = 8.0 Hz, 20H), 7.06–7.02 (m, 20H), 3.85–3.77 (m, 4H), 3.00 (dt, *J* = 9.9 Hz, *J* = 13.3 Hz, 4H).

^{13}C NMR (100 MHz, Chloroform-d [77.0 ppm], ppm) δ = 147.5 (4C), 147.5 (8C), 147.2 (4C), 144.1 (4C), 141.6 (4C), 136.6 (4C), 133.9 (4C), 132.5 (4C), 129.3 (16C), 126.6 (4C), 126.3 (8C), 124.4 (16C), 123.9 (8C), 123.1 (4C), 122.7 (8C), 33.5 (4C).

IR (ATR, ṽ) = 3055, 3023, 2929, 1585, 1547, 1482, 1455, 1313, 1268, 1190, 1176, 1112, 1074, 1024, 955, 911, 899, 867, 823, 795, 748, 722, 691, 639, 616, 509 cm^{-1}.

ESI-MS (*m/z*): 1509 (15) [M]$^+$, 756/755/754 (24/100/91) [M]$^{2+}$, 671 (24), 503 (30) [M]$^{3+}$, 316 (27), 288 (66); HRMS–ESI (*m/z*): [M]$^+$ calcd for C$_{104}$H$_{76}$N$_4$S$_4$, 1508.4953; found, 1508.4932.

4,7,13,16-Tetra(4-(2-(3,4-ethylenedioxy)thienyl)-*N*,*N*-bis(4-methoxy-phenyl)aniline)-[2.2]paracyclophane (TM_41)

4-(2-(3,4-Ethylenedioxy)thienyl)-*N*,*N*-bis(4-methoxyphenyl)aniline (14) (612 mg, 1.37 mmol, 4.80 equiv) was added to a crimp cap vial and flushed with argon three times. 6 mL of dry tetrahydrofuran was added, and the mixture was cooled to -78 °C. *n*-Butyllithium solution (96.8 mg, 605 µL, 1.51 mmol, 2.50M, 5.28 equiv) was added dropwise, and the reaction mixture was warmed to 22 °C. After stirring for 30 min, the reaction mixture was cooled to 0 °C, and zinc chloride (206 mg, 1.51 mL, 1.51 mmol, 1.00M, 5.28 equiv) in tetrahydrofuran was added. The reaction was stirred for 15 min at 0 °C and then warmed to 22 °C. 4,7,13,16-Tetrabromo[2.2]paracyclophane (150 mg, 286 µmol, 1.00 equiv) and PEPPSI™-iPr (15.6 mg, 22.9 µmol, 0.0800 equiv) was added to a crimp cap vial and flushed with argon three times. 4 mL of dry tetrahydrofuran was added, and zinc chloride (104 mg, 764 µL, 764 µmol, 1.00M, 4.00 equiv) in tetrahydrofuran was added. After dissolving, the NEGISHI reactant was added *via* a syringe, and the reaction mixture was heated to 70 °C. The reaction progress was monitored *via* TLC. After 18 h, the reaction was cooled to 22 °C, filtered through a pad of celite, and the solvent was removed under reduced pressure. The crude product was purified *via* column chromatography on basic alox (eluent cyclo-hexane/dichloromethane/ethyl acetate 20:1:1 → 4:1:1) to yield 4,7,13,16-tetra(4-(2-(3,4-ethylenedioxy)thienyl)-*N*,*N*-bis(4-methoxyphenyl)aniline)-[2.2]paracyclophane (70.0 mg, 35.3 µmol, 12% yield) as a yellow to orange solid.

M.p.: >200 °C.

R_f = 0.05 (cyclohexane/dichloromethane/ethyl acetate 2:1:1).

^{1}H NMR (400 MHz, Dichloromethane-d2 [5.32 ppm], ppm) δ = 7.61–7.59 (m, 8H), 7.08–7.04 (m, 20H), 6.88–6.81 (m, 24H), 4.26–4.18 (m, 16H), 3.78–3.76 (m, 24H), 3.60–3.53 (m, 4H), 3.06–2.98 (m, 4H).

^{13}C NMR (100 MHz, Dichloromethane-d2 [54.0 ppm], ppm) δ = 156.6 (8C), 127.0, 121.3, 115.2, 65.3 (4C), 65.1 (4C), 56.0 (8C), 27.5 (4C), due to rotational barriers in structure the resolution of ^{13}C-NMR is not good (important signals of the structure are visible with the expected shifts, 89C).

IR (ATR, ṽ) = 2925, 2866, 2861, 2850, 2832, 1604, 1585, 1502, 1462, 1451, 1436, 1360, 1317, 1276, 1237, 1194, 1179, 1135, 1085, 1034, 959, 932, 914, 874, 857, 824, 781, 722, 680, 601, 574, 555 cm^{-1}.

ESI-MS (*m/z*): 1983/1982 (1/1) [M]$^+$, 993/992/991/990 (3/30/69/49) [M]$^{2+}$, 890/889/888/887 (15/55/100/3), 661/660 (20/17); HRMS–ESI (*m/z*): [M]$^+$ calcd for $C_{120}H_{100}N_4O_{16}S_4$, 1980.6017; found, 1980.6005.

4,7,13,16-Tetra(4-(2-thienyl)-*N,N*-bis(4-*tert*-butyl-phenyl)aniline)[2.2]paracyclophane (TM_42)

4-(2-Thienyl)-*N,N*-bis(4-*tert*-butyl-phenyl)-aniline (22) (604 mg, 1.37 mmol, 4.80 equiv) was added to a crimp cap vial and flushed with argon three times. 5 mL of dry tetra-hydrofuran was added, and the mixture was cooled to -78 °C. *n*-Butyllithium solution (96.8 mg, 605 µL, 1.51 mmol, 2.50M, 5.28 equiv) was added dropwise, and the reaction mixture was warmed to 22 °C. After stirring for 30 min, the reaction mixture was cooled to 0 °C, and zinc chloride (206 mg, 1.51 mL, 1.51 mmol, 1.00M, 5.28 equiv) in tetra-hydrofuran was added. The reaction was stirred for 15 min at 0 °C and then warmed to 22 °C. 4,7,13,16-Tetrabromo[2.2]paracyclophane (150 mg, 286 µmol, 1.00 equiv) and PEPPSI™-iPr (15.6 mg, 22.9 µmol, 0.0800 equiv) was added to a crimp cap vial and flushed with argon three times. 5 mL of dry tetra-hydrofuran and zinc chloride (156 mg, 1.15 mL, 1.15 mmol, 1.00M, 4.00 equiv) in tetrahydrofuran were added. After dissolving, the NEGISHI reactant was added *via* a syringe, and the reaction mixture was heated to 70 °C. The reaction progress was monitored *via* TLC. After 18 h, the reaction was cooled to 22 °C, filtered through a pad of celite, and the solvent was removed under reduced pressure. The crude mixture was purified *via* column chromatography on silica gel (eluent cyclohexane/dichloromethane 50:1 → 5:1) to yield 4,7,13,16-tetra(4-(2-thienyl)-*N,N*-bis(4-*tert*-butylphenyl)aniline)[2.2]paracyclophane (419 mg, 214 µmol, 75% yield) as a yellow solid.

M.p.: >200 °C.

R_f = 0.3 (cyclohexane/ethyl acetate 80:1).

^{1}H NMR (400 MHz, Dichloromethane-d2 [5.32 ppm], ppm) δ = 7.48 (d, J = 8.6 Hz, 8H), 7.30–7.27 (m, 20H), 7.16 (d, J = 3.8 Hz, 4H), 7.03 (d, J = 8.8 Hz, 20H), 6.94 (d, J = 8.5 Hz, 8H), 3.81–3.77 (m, 4H), 3.02–2.98 (m, 4H), 1.30 (s, 72H).

^{13}C NMR (100 MHz, Dichloromethane-d2 [54.0 ppm], ppm) δ = 148.3 (4C), 146.6 (8C), 145.4 (8C), 144.8 (4C), 142.0 (4C), 137.2 (4C), 134.4 (4C), 133.1 (4C), 128.1 (4C), 127.2 (4C), 126.7 (16C), 126.6 (8C), 124.7 (16C), 123.5 (8C), 123.0 (4C), 34.8 (4C), 34.0 (24C), 31.8 (8C).

IR (ATR, ṽ) = 3031, 2952, 2925, 2901, 2863, 1598, 1506, 1483, 1459, 1391, 1363, 1319, 1292, 1265, 1201, 1184, 1109, 1014, 912, 827, 796, 737, 724, 700, 581, 568, 552 cm^{-1}.

ESI-MS (m/z): 1958 (5) [M]$^+$, 981/980/979/978 (17/83/100/59) [M]$^{2+}$, 653/652 (39/26) [M]$^{3+}$, 283/282 (9/44); HRMS–ESI (m/z): [M]$^+$ calcd for $C_{136}H_{140}N_4S_4$, 1956.9961; found, 1956.9981.

4,7,13,16-Tetra(9-(4-(2-thienyl)phenyl)-3,6-dimethoxycarb-azole)[2.2]paracyclophane (TM_43)

9-(4-(2-Thienyl)phenyl)-3,6-di-methoxycarbazole (25) (530 mg, 1.37 mmol, 4.80 equiv) was added to a crimp cap vial and flushed with argon three times. 5 mL of dry tetrahydro-furan was added, and the mixture was cooled to -78 °C. n-Butyllithium solution (96.8 mg, 605 μL, 1.51 mmol, 2.50M, 5.28 equiv) was added dropwise, and the reaction mixture was warmed to 22 °C. After stirring for 30 min, the reaction mixture was cooled to 0 °C, and zinc chloride (206 mg, 1.51 mL, 1.51 mmol, 1.00M, 5.28 equiv) in tetrahydrofuran was added. The reaction was stirred for 15 min at 0 °C and then warmed to 22 °C. 4,7,13,16-Tetrabromo[2.2]paracyclophane (150 mg, 286 μmol, 1.00 equiv) and PEPPSI™-iPr (15.6 mg, 22.9 μmol, 0.0800 equiv) was added to a crimp cap vial and flushed with argon three times. 5 mL of dry tetrahydrofuran and zinc chloride (156 mg, 1.15 mL, 1.15 mmol, 1.00M, 4.00 equiv) in tetrahydrofuran were added. After dissolving, the NEGISHI reactant was added *via* a syringe, and the reaction mixture was heated to 70 °C. The reaction progress was monitored *via* TLC. After 18 h, the reaction was cooled to 22 °C, filtered through a pad of celite, and the solvent was removed under reduced pressure. The crude mixture was purified *via* column chromatography on silica gel (eluent cyclo-hexane/dichloromethane/ethyl acetate 4:1:0 → 0:4:1) to yield 4,7,13,16-tetra(9-(4-(2-thienyl)phenyl)-3,6-dimethoxycarbazole)[2.2]paracyclophane (277 mg, 159 μmol, 56% yield) as a yellow solid.

M.p.: >200 °C.

R_f = 0.45 (cyclohexane/ethyl acetate 1:1).

^{1}H NMR (400 MHz, Chloroform-d [7.27 ppm], ppm) δ = 7.88 (d, J = 8.3 Hz, 8H), 7.53–7.48 (m, 20H), 7.29 (d, J = 9.8 Hz, 12H), 7.16 (s, 4H), 6.89 (dd, J = 2.1 Hz, J = 8.8 Hz, 8H), 3.97–3.96 (m, 4H), 3.91 (s, 24H), 3.17–3.09 (m, 4H).

^{13}C NMR (100 MHz, Chloroform-d [77.0 ppm], ppm) δ = 154.0 (8C), 143.3 (4C), 143.0 (4C), 137.4 (4C), 136.8 (4C), 135.8 (8C), 134.0 (4C), 132.9 (4C), 132.5 (4C), 126.9 (4C), 126.8 (8C), 126.6 (8C), 124.0 (4C), 123.8 (8C), 114.8 (8C), 110.4 (8C), 103.1 (8C), 56.0 (8C), 33.5 (4C).

IR (ATR, ṽ) = 2927, 2898, 2873, 2853, 2823, 1737, 1605, 1581, 1538, 1511, 1487, 1462, 1431, 1384, 1367, 1327, 1306, 1283, 1261, 1201, 1174, 1153, 1106, 1033, 980, 959, 946, 909, 899, 867, 854, 833, 819, 790, 751, 724, 708, 687, 660, 633, 595, 584 cm^{-1}.

ESI-MS (*m/z*): 1742/1741 (1/1) [M]$^+$, 873/872/871/870 (1/9/42/34) [M]$^{2+}$, 582/581/580 (1/25/18) [M]$^{3+}$, 463/462 (28/100), 308/307/306 (19/5/29), 297 (21), 282 (39); HRMS–ESI (*m/z*): [M]$^+$ calcd for $C_{112}H_{84}N_4O_8S_4$, 1740.5172; found, 1740.5185.

5.2.4 *Meso*-tetrakis-substituted Porphyrins

Meso-tetrakis-(4-(2-thienyl)-*N*,*N*-bis(4-methoxyphenyl)aniline)porphyrin-Zn(II) (TM_44_Zn)

4-(2-Thienyl)-*N*,*N*-bis(4-methoxyphenyl)-aniline (12) (500 mg, 1.29 mmol, 4.80 equiv) was added to a crimp cap vial and flushed with argon three times. 5 mL of dry tetrahydrofuran was added, and the mixture was cooled to -78 °C. *n*-Butyllithium solution (90.9 mg, 568 µL, 1.42 mmol, 2.50M, 5.28 equiv) was added dropwise, and the reaction mixture was warmed to 22 °C. After stirring for 30 min, the reaction mixture was cooled to 0 °C, and zinc chloride (193 mg, 1.42 mL, 1.42 mmol, 1.00M, 5.28 equiv) in tetrahydrofuran was added. The reaction was stirred for 15 min at 0 °C and then warmed to 22 °C. *Meso*-tetrakis-(4-bromophenyl)porphyrin (250 mg, 269 µmol, 1.00 equiv) and PEPPSI™-iPr (14.7 mg, 21.5 µmol, 0.0800 equiv) were added to a crimp cap vial and flushed with argon three times. 5 mL of dry tetrahydrofuran was added, and zinc chloride (146 mg, 1.07 mL, 1.07 mmol, 1.00M, 4.00 equiv) in tetra-hydrofuran was added. After dissolving, the NEGISHI reactant was added *via* a syringe, and the reaction mixture was heated to 70 °C. The reaction progress was monitored *via* TLC. After 19 h, the reaction was cooled to 22 °C, and the solvent was removed under reduced pressure. The crude product was purified *via* column chromatography on silica gel (eluent cyclohexane/tetrahydrofuran 3:1 → 0:1). After evaporating the solvents, the crude product was washed with methanol to yield *meso*-tetrakis-(4-(2-thienyl)-*N*,*N*-bis(4-methoxy-phenyl)aniline))porphyrin-Zn(II) (410 mg, 185 µmol, 69% yield) as a red solid.

M.p.: >300 °C.

R_f = 0.1 (cyclohexane/tetrahydrofuran 3:1).

^{1}H NMR (500 MHz, THF-d8 [3.58 ppm], ppm) δ = 8.97 (s, 8H), 8.23 (d, J = 8.1 Hz, 8H), 8.06 (d, J = 8.1 Hz, 8H), 7.66 (d, J = 3.8 Hz, 4H), 7.56 (d, J = 8.7 Hz, 8H), 7.38 (d, J = 3.8 Hz, 4H), 7.08–7.07 (m, 16H), 6.94 (d, J = 8.7 Hz, 8H), 6.87 (d, J = 9.0 Hz, 16H), 3.77 (s, 24H).

^{13}C NMR (125 MHz, THF-d8 [67.6 ppm], ppm) δ = 157.6 (8C), 151.2 (8C), 149.7 (4C), 145.3 (4C), 143.6 (4C), 142.8 (4C), 141.7 (8C), 136.1 (8C), 134.7 (4C), 132.5 (8C), 127.7 (16C), 127.5 (4C), 127.1 (8C), 125.7 (4C), 124.3 (4C), 123.9 (8C), 121.4 (8C), 121.3 (4C), 115.7 (16C), 55.8 (8C).

IR (ATR, ṽ) = 2945, 2927, 2901, 2829, 1599, 1496, 1453, 1439, 1317, 1275, 1235, 1191, 1177, 1164, 1103, 1067, 1033, 996, 972, 939, 912, 824, 793, 718, 700, 646, 632, 599, 575, 526 cm^{-1}.

ESI-MS (m/z): 2218 [M]$^+$, 1108 [M]$^{2+}$; HRMS–ESI (m/z): [M]$^+$ calcd for $C_{140}H_{104}N_8O_8S_4Zn_1$, 2218.62; found, 2218.62.

Meso-tetrakis-(4-(2-(3,4-ethylenedioxy)thienyl)-*N*,*N*-bis(4-methoxy-phenyl)aniline)porphyrin-Zn(II) (TM_45_Zn)

4-(2-(3,4-Ethylenedi-oxy)thienyl)-*N*,*N*-bis(4-methoxyphenyl)aniline (14) (460 mg, 1.03 mmol, 4.80 equiv) was added to a crimp cap vial and flushed with argon three times. 5 mL of dry tetra-hydrofuran was added, and the mixture was cooled to -78 °C. *n*-Butyllithium solution (72.7 mg, 454 μL, 1.14 mmol, 2.50M, 5.28 equiv) was added dropwise, and the reaction mixture was warmed to 22 °C. After stirring for 30 min, the reaction mixture was cooled to 0 °C, and zinc chloride (155 mg, 1.14 mL, 1.14 mmol, 1.00M, 5.28 equiv) in tetrahydrofuran was added. The reaction was stirred for 15 min at 0 °C and then warmed to 22 °C. *Meso*-tetrakis-(4-bromophenyl)porphyrin (200 mg, 215 μmol, 1.00 equiv) and PEPPSI™-iPr (11.7 mg, 17.2 μmol, 0.0800 equiv) were added to a crimp cap vial and flushed with argon three times. 5 mL of dry tetrahydrofuran was added, and zinc chloride (117 mg, 860 μL, 860 μmol, 1.00M, 4.00 equiv) in tetrahydrofuran was added. After dissolving, the NEGISHI reactant was added *via* a syringe, and the reaction mixture was heated to 70 °C. The reaction progress was monitored *via* TLC. After 19 h, the reaction was cooled to 22 °C, and the solvent was removed under reduced pressure. The crude mixture was purified *via* column chromatography on silica gel (eluent cyclohexane/tetrahydrofuran 2:1 → 0:1). After evaporating the solvent under reduced pressure, the crude product was washed with methanol to yield *meso*-tetrakis-(4-(2-(3,4-ethylenedioxy)thienyl)-*N*,*N*-bis(4-methoxy-phenyl)aniline))porphyrin-Zn(II) (380 mg, 155 μmol, 72% yield) as a red solid.

M.p.: >300 °C.

R_f = 0.1 (cyclohexane/tetrahydrofuran 3:1).

^{1}H NMR (500 MHz, THF-d8 [3.58 ppm], ppm) δ = 8.96 (s, 8H), 8.18 (dd, J = 8.2 Hz, J = 18.3 Hz, 16H), 7.66 (d, J = 8.9 Hz, 8H), 7.07–7.04 (m, 16H), 6.93 (d, J = 8.9 Hz, 8H), 6.86–6.84 (m, 16H), 4.46 (dd, J = 29.3 Hz, J = 3.5 Hz, 16H), 3.76 (s, 24H).

^{13}C NMR (125 MHz, THF-d8 [67.6 ppm], ppm) δ = 157.4 (8C), 151.2 (8C), 148.6 (4C), 142.5 (4C), 141.9 (8C), 140.8 (8C), 139.3 (8C), 135.8 (8C), 133.6 (4C), 132.5 (4C), 127.7 (8C), 127.5 (16C), 126.6 (4C), 124.6 (8C), 121.5 (8C), 116.9 (4C), 115.6 (16C), 114.4 (4C), 66.0 (4C), 65.8 (4C), 55.8 (8C).

IR (ATR, $\tilde{\nu}$) = 2921, 2830, 1598, 1588, 1499, 1466, 1459, 1451, 1434, 1401, 1357, 1316, 1279, 1232, 1193, 1176, 1163, 1132, 1081, 1030, 994, 966, 911, 823, 807, 795, 718, 680, 633, 589, 569, 548, 513 cm^{-1}.

ESI-MS (m/z): 1634 [(M)$_2$H]$^+$, 1225 [M]$^{2+}$, 816 [M]$^{3+}$, 612 [M]$^{4+}$; HRMS–ESI (m/z): [M]$^{2+}$ calcd for $C_{148}H_{112}N_8O_{16}S_4Zn_1$, 1225.32; found, 1225.32.

Meso-tetrakis-(9-(4-(2-thienyl)phenyl)-3,6-dimethoxycarbazole)porphyrin-Zn(II) (TM_46_Zn)

9-(4-(2-Thienyl)phenyl)-3,6-dimethoxycarbazole (25) (497 mg, 1.29 mmol, 4.80 equiv) was added to a crimp cap vial and flushed with argon three times. 5 mL of dry tetrahydrofuran was added, and the mixture was cooled to -78 °C. *n*-Butyllithium solution (90.9 mg, 568 µL, 1.42 mmol, 2.50M, 5.28 equiv) was added dropwise, and the reaction mixture was warmed to 22 °C. After stirring for 30 min, the reaction mixture was cooled to 0 °C, and zinc chloride (193 mg, 1.42 mL, 1.42 mmol, 1.00M, 5.28 equiv) in tetrahydrofuran was added. The reaction was stirred for 15 min at 0 °C and then warmed to 22 °C. *Meso*-tetrakis-(4-bromophenyl)porphyrin (250 mg, 269 µmol, 1.00 equiv) and PEPPSI™-iPr (14.7 mg, 21.5 µmol, 0.0800 equiv) were added to a crimp cap vial and flushed with argon three times. 5 mL of dry tetrahydrofuran was added, and zinc chloride (146 mg, 1.07 mL, 1.07 mmol, 1.00M, 4.00 equiv) in tetrahydrofuran was added. After dissolving, the NEGISHI reactant was added *via* a syringe, and the reaction mixture was heated to 70 °C. The reaction progress was monitored *via* TLC. After 19 h, the reaction was cooled to 22 °C, filtered through a pad of celite, and the solvent was removed under reduced pressure. The crude product was purified *via* column chromatography on silica gel (eluent cyclohexane/tetrahydrofuran 3:1 → 1:2), and after evaporating the solvents the crude product was washed with methanol to yield *meso*-tetrakis-(9-(4-(2-thienyl)phenyl)-3,6-dimethoxycarbazole))porphyrin-Zn(II) (510 mg, 231 µmol, 86% yield) as a red solid.

M.p.: >300 °C.

R_f = 0.1 (cyclohexane/tetrahydrofuran 3:1).

^{1}H NMR (500 MHz, THF-d8 [3.58 ppm], ppm) δ = 9.01 (s, 8H), 8.28–8.27 (m, 8H), 8.11 (d, J = 7.8 Hz, 8H), 8.01 (d, J = 8.2 Hz, 8H), 7.73 (d, J = 3.7 Hz, 4H), 7.68–7.63 (m, 20H), 7.42 (d, J = 8.9 Hz, 8H), 7.03 (dd, J = 2.3 Hz, J = 8.9 Hz, 8H), 3.90 (s, 24H).

^{13}C NMR (125 MHz, THF-d8 [67.6 ppm], ppm) δ = 155.7 (8C), 151.2 (8C), 144.7 (4C), 144.0 (8C), 138.8 (4C), 137.1 (8C), 136.2 (8C), 134.5 (4C), 133.9 (4C), 132.6 (8C), 127.9 (8C), 127.9 (4C), 127.8 (8C), 126.0 (4C), 125.9 (8C), 125.2 (4C), 124.5 (4C), 121.3 (4C), 116.1 (8C), 111.5 (8C), 103.8 (8C), 56.2 (8C).

IR (ATR, ṽ) = 2928, 2825, 1602, 1581, 1516, 1486, 1462, 1429, 1364, 1326, 1303, 1281, 1259, 1200, 1173, 1152, 1105, 1067, 1030, 996, 972, 946, 911, 832, 789, 752, 738, 718, 659, 584 cm^{-1}.

ESI-MS (m/z): 2211 [M+H]$^+$; HRMS–ESI (m/z): [M+H]$^+$ calcd for $C_{140}H_{97}N_8O_8S_4Zn_1$, 2211.56; found, 2211.57.

Meso-tetrakis-(9-(4-(2-(3,4-ethylenedioxy)thienyl)phenyl)-3,6-dimethoxy-carbazole)porphyrin-Zn(II) (TM_47_Zn)

9-(4-(2-(3,4-Ethylenedi-oxy)thienyl)phenyl)-3,6-dimethoxycarbazole (29) (572 mg, 1.29 mmol, 4.80 equiv) was added to a crimp cap vial and flushed with argon three times. 5 mL of dry tetrahydrofuran was added, and the mixture was cooled to -78 °C. *n*-Butyllithium solution (90.9 mg, 568 µL, 1.42 mmol, 2.50M, 5.28 equiv) was added dropwise, and the reaction mixture was warmed to 22 °C. After stirring for 30 min, the reaction mixture was cooled to 0 °C, and zinc chloride (193 mg, 1.42 mL, 1.42 mmol, 1.00M, 5.28 equiv) in tetrahydrofuran was added. The reaction was stirred for 15 min at 0 °C and then warmed to 22 °C. *Meso*-tetrakis-(4-bromophenyl)porphyrin (250 mg, 269 µmol, 1.00 equiv) and PEPPSI™-iPr (14.7 mg, 21.5 µmol, 0.0800 equiv) were added to a crimp cap vial and flushed with argon three times. 5 mL of dry tetrahydrofuran was added, and zinc chloride (146 mg, 1.07 mL, 1.07 mmol, 1.00M, 4.00 equiv) in tetrahydrofuran was added. After dissolving, the NEGISHI reactant was added *via* a syringe, and the reaction mixture was heated to 70 °C. The reaction progress was monitored *via* TLC. After 21 h, the reaction was cooled to 22 °C, filtered through a pad of celite, and the solvent was removed under reduced pressure. The crude product was purified *via* column chromatography on silica gel (eluent cyclohexane/tetrahydrofuran 2:1 → 1:4). After evaporation of the solvents, the crude product was washed with methanol to yield *meso*-tetrakis-(9-(4-(2-(3,4-ethylenedioxy)thienyl)phenyl)-3,6-dimethoxycarbazole))porphyrin-Zn(II) (205 mg, 83.9 µmol, 31% yield) as a red solid.

M.p.: >300 °C.

R_f = 0.1 (cyclohexane/tetrahydrofuran 3:1).

^{1}H NMR (500 MHz, THF-d8 [3.58 ppm], ppm) δ = 9.00 (s, 8H), 8.27–8.23 (m, 16H), 8.12 (d, J = 8.4 Hz, 8H), 7.66 (d, J = 2.3 Hz, 8H), 7.64 (d, J = 8.5 Hz, 8H), 7.41 (d, J = 8.9 Hz, 8H), 7.03 (dd, J = 2.4 Hz, J = 9.0 Hz, 8H), 4.56 (dd, J = 4.9 Hz, J = 12.7 Hz, 16H), 3.90 (s, 24H).

^{13}C NMR (125 MHz, THF-d8 [67.6 ppm], ppm) δ = 155.6 (8C), 151.2 (8C), 143.0 (4C), 140.9 (4C), 140.8 (8C), 137.7 (4C), 137.1 (8C), 135.9 (4C), 133.4 (4C), 132.9 (8C), 132.5 (4C), 128.2 (8C), 127.6 (8C), 125.1(8C), 124.9 (8C), 121.5 (4C), 116.3 (4C), 116.0 (8C), 115.6 (4C), 111.5 (8C), 103.8 (8C), 66.0 (8C), 56.2 (8C).

IR (ATR, ṽ) = 2929, 2917, 2897, 2881, 2873, 2825, 1581, 1520, 1506, 1487, 1462, 1429, 1404, 1357, 1327, 1306, 1285, 1261, 1201, 1173, 1152, 1135, 1108, 1081, 1030, 996, 967, 948, 912, 867, 853, 832, 789, 754, 725, 717, 694, 657, 636, 598, 585, 564, 548, 535 cm^{-1}.

ESI-MS (m/z): 1221 [M]$^{2+}$, 814 [M]$^{3+}$, 610 [M]$^{4+}$; HRMS–ESI (m/z): [M]$^{2+}$ calcd for $C_{148}H_{104}N_8O_{16}S_4Zn_1$, 1221.29; found, 1221.29.

Meso-tetrakis-(4-(2-selenyl)-*N,N*-bis(4-methoxyphenyl)aniline)porphyrin-Zn(II) (TM_48_Zn)

4-(2-Selenyl)-*N,N*-bis(4-methoxyphenyl)aniline (30) (448 mg, 1.03 mmol, 4.80 equiv) was added to a crimp cap vial and flushed with argon three times. 5 mL of dry tetrahydrofuran was added, and the mixture was cooled to -78 °C. *n*-Butyllithium solution (72.7 mg, 454 µL, 1.14 mmol, 2.50M, 5.28 equiv) was added dropwise, and the reaction mixture was warmed to 22 °C. After stirring for 30 min, the reaction mixture was cooled to 0 °C, and zinc chloride (155 mg, 1.14 mL, 1.14 mmol, 1.00M, 5.28 equiv) in tetrahydrofuran was added. The reaction was stirred for 15 min at 0 °C and then warmed to 22 °C. *Meso*-tetrakis-(4-bromophenyl)porphyrin (200 mg, 215 µmol, 1.00 equiv) and PEPPSI™-iPr (11.7 mg, 17.2 µmol, 0.0800 equiv) were added to a crimp cap vial and flushed with argon three times. 5 mL of dry tetrahydrofuran was added, and zinc chloride (117 mg, 860 µL, 860 µmol, 1.00M, 4.00 equiv) in tetrahydrofuran was added. After dissolving, the NEGISHI reactant was added *via* a syringe, and the reaction mixture was heated to 70 °C. The reaction progress was monitored *via* TLC. After 18 h, the reaction was cooled to 22 °C, and the solvent was removed under reduced pressure. The crude mixture was purified *via* column chromatography on silica gel (eluent cyclohexane/tetrahydrofuran 1:1 → 0:1 + 10% pyridine) to yield *meso*-tetrakis-(4-(2-selenyl)-*N,N*-bis(4-methoxyphenyl)aniline)porphyrin-Zn(II) (180 mg, 74.8 µmol, 35% yield) as a dark red solid.

M.p.: >300 °C

R_f = 0.1 (cyclohexane/tetrahydrofuran 3:1).

^{1}H NMR (500 MHz, THF-d8 [3.58 ppm], pyridine-d5, ppm) δ = 8.96 (s, 8H), 8.19 (d, J = 7.8 Hz, 8H), 7.97 (d, J = 8.1 Hz, 8H), 7.79 (d, J = 3.7 Hz, 4H), 7.52–7.49 (m, 12H), 7.08 (d, J = 8.9 Hz, 16H), 6.93 (d, J = 8.5 Hz, 8H), 6.88 (d, J = 8 9 Hz, 16H), 3.76 (s, 24H).

^{13}C NMR (125 MHz, THF-d8 [67.6 ppm], pyridine-d5, ppm) δ = 157.6 (8C), 151.2 (4C), 151.1 (8C), 149.8 (4C), 148.6 (4C), 143.8 (8C), 141.6 (8C), 136.6 (8C), 132.5 (8C), 129.6 (4C), 127.9 (4C), 127.7 (16C), 127.5 (8C), 125.9 (4C), 124.6 (4C), 124.2 (4C), 121.4 (8C), 121.3 (4C), 115.7 (16C), 55.8 (8C).

IR (ATR, $\tilde{v}$) = 2946, 2924, 2902, 2829, 1598, 1499, 1462, 1439, 1404, 1337, 1317, 1278, 1264, 1237, 1191, 1177, 1106, 1069, 1033, 997, 952, 911, 824, 795, 720, 686, 667, 639, 602, 575, 547 cm^{-1}.

ESI-MS (m/z): 1203 [M]$^{2+}$, 802 [M]$^{3+}$; HRMS–ESI (m/z): [M]$^{2+}$ calcd for $C_{140}H_{104}N_8O_8Se_4Zn_1$, 1203.70; found, 1203.70.

189

Meso-tetrakis-(4-(2-thienyl)-*N*,*N*-bis(4-methoxyphenyl)aniline)porphyrin (TM_44_2H)

Meso-tetrakis-(4-(2-thienyl)-*N*,*N*-bis(4-methoxy-phenyl)aniline)porphyrin-Zn(II) (TM_44_Zn) (300 mg, 135 μmol, 1.00 equiv) was dissolved in 60 mL dichloromethane, 2,2,2-trifluoroacetic acid (1.16 g, 776 μL, 10.1 mmol, 75.0 equiv) was added, and the reaction was stirred at 22 °C for 1 h. The reaction progress was monitored *via* TLC, and after full conversion, the product mixture was neutralized with saturated NaHCO$_3$ solution, and the product was extracted with dichloromethane. The combined organic phases were washed with water (3x) and brine (3x) and dried over anhydrous Na$_2$SO$_4$. After evaporating the solvent under reduced pressure, the crude product was precipitated from a cyclohexane/dichloromethane mixture and washed with cyclohexane three times to yield *meso*-tetrakis-(4-(2-thienyl)-*N*,*N*-bis(4-methoxy-phenyl)aniline)porphyrin (291 mg, 135 μmol, 100% yield) as a red solid.

M.p.: >300 °C.

R$_f$ = 0.1 (cyclohexane/tetrahydrofuran 3:1).

^{1}H NMR (500 MHz, THF-d8 [3.58 ppm], ppm) δ = 8.88 (s, 8H), 8.12 (d, *J* = 8.1 Hz, 8H), 7.97 (d, *J* = 8.1 Hz, 8H), 7.57 (d, *J* = 3.8 Hz, 4H), 7.53 (d, *J* = 8.7 Hz, 8H), 7.33 (d, *J* = 3.7 Hz, 4H), 7.05 (d, *J* = 8.9 Hz, 16H), 6.93 (d, *J* = 8.7 Hz, 8H), 6.84 (d, *J* = 8.9 Hz, 16H), 3.75 (s, 24H), -2.62 (s, 2H).

^{13}C NMR (125 MHz, THF-d8 [67.6 ppm], ppm) δ = 157.6 (8C), 149.7 (4C), 145.4 (4C), 142.5 (4C), 142.1 (4C), 141.7 (8C), 136.1 (4C), 135.1 (4C), 127.7 (16C), 127.5 (8C), 127.1 (8C), 125.8 (4C), 124.5 (8C), 123.9 (4C), 121.4 (8C), 120.8 (4C), 115.7 (16C), 55.8 (8C), missing carbons due to interaction in porphyrin (16C).

IR (ATR, ṽ) = 2924, 2900, 2827, 1735, 1725, 1598, 1560, 1493, 1452, 1439, 1400, 1347, 1316, 1275, 1235, 1190, 1177, 1163, 1103, 1031, 992, 983, 965, 911, 824, 790, 730, 721, 700, 633, 598, 575, 524, 507 cm^{-1}.

ESI-MS (m/z): 2156 [M]$^+$, 1077 [M]$^{2+}$, 718 [M]$^{3+}$; HRMS–ESI (m/z): [M]$^+$ calcd for $C_{140}H_{106}N_8O_8S_4$, 2156.71; found, 2156.71.

Meso-tetrakis-(4-(2-(3,4-ethylenedioxy)thienyl)-*N,N*-bis(4-methoxy-phenyl)aniline)porphyrin (TM_45_2H)

Meso-tetrakis-(4-(2-(3,4-ethylenedioxy)thienyl)-*N,N*-bis(4-methoxy-phenyl)aniline)porphyrin-Zn(II) (TM_45_Zn) (300 mg, 122 μmol, 1.00 equiv) was dissolved in 55 mL dichloromethane, 2,2,2-trifluoroacetic acid (1.05 g, 702 μL, 9.18 mmol, 75.0 equiv) was added, and the reaction was stirred at 22 °C for 1 h. The reaction progress was monitored *via* TLC, and after full conversion, the product mixture was neutralized with saturated NaHCO$_3$ solution, and the product was extracted with dichloromethane. The combined organic phases were washed with water (3x) and brine (3x) and dried over anhydrous Na$_2$SO$_4$. After evaporating the solvent under reduced pressure, the crude product was precipitated from a cyclohexane/dichloromethane mixture and washed with cyclohexane three times to yield *meso*-tetrakis-(4-(2-(3,4-ethylenedi-oxy)thienyl)-*N,N*-bis(4-methoxyphenyl)aniline)porphyrin (292 mg, 122 µmol, 100% yield) as a red solid.

M.p.: >300 °C.

R_f = 0.1 (cyclohexane/tetrahydrofuran 3:1).

^{1}H NMR (500 MHz, THF-d8 [3.58 ppm], ppm) δ = 8.93 (s, 8H), 8.18 (d, J = 1.7 Hz, 16H), 7.65 (d, J = 8.9 Hz, 8H), 7.04 (d, J = 8.9 Hz, 16H), 6.92 (d, J = 8.7 Hz, 8H), 6.83 (s, 16H), 4.45 (dd, J = 3.3 Hz, J = 30.3 Hz, 16H), 3.75 (s, 24H), -2.60 (s, 2H).

^{13}C NMR (125 MHz, THF-d8 [67.6 ppm], ppm) δ = 157.4 (8C), 148.7 (4C), 141.8 (8C), 141.1 (4C), 141.0 (4C), 139.3 (4C), 135.8 (8C), 134.1 (4C), 127.7 (8C), 127.5 (16C), 126.5 (4C), 124.9 (8C), 121.4 (8C), 121.0 (4C), 117.2 (4C), 115.6 (16C), 114.2 (4C), 66.0 (4C), 65.8 (4C), 55.8 (8C), missing carbons due to interaction in porphyrin (16C).

IR (ATR, ṽ) = 2927, 2829, 1599, 1587, 1499, 1463, 1452, 1435, 1401, 1357, 1316, 1281, 1234, 1193, 1176, 1163, 1133, 1106, 1081, 1031, 992, 983, 965, 912, 853, 823, 798, 721, 680, 591, 572, 550 cm^{-1}.

ESI-MS (*m/z*): 1193 [M]$^{2+}$, 795 [M]$^{3+}$, 596 [M]$^{4+}$; HRMS–ESI (*m/z*): [M]$^{2+}$ calcd for $C_{148}H_{114}N_8O_{16}S_4$, 1193.86; found, 1193.87.

Meso-tetrakis-(9-(4-(2-thienyl)phenyl)-3,6-dimethoxycarbazole)porphyrin (TM_46_2H)

Meso-tetrakis-(9-(4-(2-thienyl)phenyl)-3,6-dimethoxy-carbazole)porphyrin-Zn(II) (TM_46_Zn) (300 mg, 136 µmol, 1.00 equiv) was dissolved in 60 mL dichloromethane, 2,2,2-tri-fluoroacetic acid (1.16 g, 778 µL, 10.2 mmol, 75.0 equiv) was added, and the reaction was stirred at 22 °C for 1 h. The reaction progress was monitored *via* TLC, and after full conversion, the product mixture was neutralized with saturated NaHCO$_3$ solution, and the product was extracted with di-chloromethane. The combined organic phases were washed with water (3x) and brine (3x) and dried over anhydrous Na$_2$SO$_4$. After evaporating the solvent under reduced pressure, the crude product was precipitated from a cyclohexane/dichloromethane mixture and washed with cyclohexane three times to yield *meso*-tetrakis-(9-(4-(2-thienyl)phenyl)-3,6-dimethoxycarb-azole)porphyrin (291 mg, 135 µmol, 100% yield) as a red solid.

M.p.: >300 °C.

R$_f$ = 0.1 (cyclohexane/tetrahydrofuran 3:1).

^{1}H NMR (500 MHz, THF-d8 [3.58 ppm], ppm) δ = 8.90 (s, 8H), 8.15 (s, 8H), 8.01 (s, 4H), 7.95 (d, J = 5.8 Hz, 8H), 7.65–7.63 (m, 20H), 7.55 (s, 8H), 7.39 (d, J = 9.3 Hz, 8H), 7.00 (d, J = 8.1 Hz, 8H), 3.88 (s, 24H), -2.59 (s, 2H).

^{13}C NMR (125 MHz, THF-d8 [67.6 ppm], ppm) δ = 155.7 (8C), 144.4 (4C), 144.1 (4C), 142.5 (4C), 138.8 (4C), 137.0 (8C), 136.2 (8C), 134.8 (4C), 133.8

(8C), 127.9 (8C), 127.8 (8C), 126.1 (4C), 125.9 (8C), 125.2 (4C), 124.8 (4C), 120.8 (4C), 116.0 (8C), 111.5 (8C), 103.8 (8C), 56.2 (8C), missing carbons due to interaction in porphyrin (16C).

IR (ATR, ṽ) = 2919, 2895, 2825, 1601, 1579, 1487, 1462, 1429, 1364, 1349, 1326, 1303, 1281, 1258, 1200, 1173, 1152, 1105, 1030, 992, 983, 965, 946, 911, 898, 832, 788, 724, 659, 632, 585, 571 cm^{-1}.

ESI-MS (m/z): 2148 [M+H]$^+$; HRMS–ESI (m/z): [M+H]$^+$ calcd for C$_{140}$H$_{99}$N$_8$O$_8$S$_4$, 2148.65; found, 2148.65.

Meso-tetrakis-(9-(4-(2-(3,4-ethylenedioxy)thienyl)phenyl)-3,6-dimethoxy-carbazole)porphyrin (TM_47_2H)

Meso-tetrakis-(9-(4-(2-(3,4-ethylenedioxy)thienyl)phenyl)-3,6-dimethoxycarbazole)-porphyrin-Zn(II) (TM_47_Zn) (200 mg, 81.8 µmol, 1.00 equiv) was dissolved in 50 mL dichloromethane and 2,2,2-trifluoroacetic acid (700 mg, 470 µL, 6.14 mmol, 75.0 equiv) was added. The reaction mixture was stirred at 22 °C for 1 h, and the reaction progress was monitored *via* TLC. The product mixture was neutralized with saturated NaHCO$_3$ solution, and the product was extracted with dichloromethane. The combined organic phases were washed with water (3x) and brine (3x) and dried over anhydrous Na$_2$SO$_4$. After evaporating the solvent under reduced pressure, the crude product was precipitated from a cyclohexane/dichloromethane mixture and washed with cyclohexane three times to yield *meso*-tetrakis-(9-(4-(2-(3,4-ethylenedioxy)thienyl)phenyl)-

3,6-dimethoxycarbazole)porphyrin (192 mg, 80.6 μmol, 99% yield) as a red solid.

M.p.: >300 °C.

R_f = 0.2 (cyclohexane/tetrahydrofuran 3:1).

^{1}H NMR (500 MHz, THF-d8 [3.58 ppm], ppm) δ = 8.96 (s, 8H), 8.23 (s, 16H), 8.10 (d, J = 8.5 Hz, 8H), 7.65 (d, J = 2.3 Hz, 8H), 7.62 (d, J = 8.4 Hz, 8H), 7.40 (d, J = 8.9 Hz, 8H), 7.01 (dd, J = 2.4 Hz, J = 8.9 Hz, 8H), 4.53 (d, J = 8.4 Hz, 16H), 3.89 (s, 24H), -2.58 (s, 2H).

^{13}C NMR (125 MHz, THF-d8 [67.6 ppm], ppm) δ = 155.6 (8C), 141.6 (4C), 141.0 (8C), 140.8 (8C), 137.7 (4C), 137.1 (8C), 135.9 (4C), 133.9 (4C), 132.8 (4C), 128.2 (8C), 127.6 (8C), 125.2 (4C), 125.1 (8C), 121.0 (4C), 116.1 (4C), 116.0 (8C), 115.8 (4C), 111.5 (8C), 103.8 (8C), 66.0 (8C), 56.2 (8C), missing carbons due to interaction in porphyrin (16C).

IR (ATR, ṽ) = 2928, 2924, 2917, 2902, 2897, 2881, 2871, 2863, 2825, 1725, 1601, 1581, 1523, 1506, 1487, 1460, 1431, 1402, 1357, 1326, 1307, 1285, 1261, 1200, 1173, 1152, 1135, 1108, 1081, 1030, 992, 983, 965, 948, 911, 853, 832, 788, 725, 688, 657, 635, 596, 585, 562 cm^{-1}.

ESI-MS (m/z): 2380 [M]$^+$, 1190 [M]$^{2+}$, 793 [M]$^{3+}$; HRMS–ESI (m/z): [M]+ calcd for $C_{148}H_{106}N_8O_{16}S_4$, 2380.67; found, 2380.67.

Meso-tetrakis-(4-(2-thienyl)-*N*,*N*-bis(4-methoxyphenyl)aniline)porphyrin-Ni(II) (TM_44_Ni)

Meso-tetrakis-(4-(2-thienyl)-*N*,*N*-bis(4-methoxy-phenyl)aniline)porphyrin (TM_44_2H) (100 mg, 46.4 μmol, 1.00 equiv) and Ni(OAc)$_2$*4 H$_2$O (57.7 mg, 232 μmol, 5.00 equiv) was added to a 20 mL crimp cap vial and 12 mL of *N*,*N*-dimethylformamide was added. The reaction was heated to 140 °C for 22 h, and the reaction progress was monitored *via* TLC. The product mixture was cooled to 22 °C, and the solvent was removed under reduced pressure. The crude product was purified *via* column chromatography on silica gel (eluent cyclohexane/tetrahydrofuran 3:1 → 0:1) and after evaporation of the solvent under reduced pressure washed with methanol to yield *meso*-tetrakis-(4-(2-thienyl)-*N*,*N*-bis(4-methoxyphenyl)aniline)porphyrin-Ni(II) (77.0 mg, 34.8 μmol, 75% yield) as a red solid.

M.p.: >300 °C.

R$_f$ = 0.15 (cyclohexane/tetrahydrofuran).

^{1}H NMR (500 MHz, THF-d8 [3.58 ppm], ppm) δ = 8.79 (s, 8H), 7.98–7.95 (m, 16H), 7.57 (d, *J* = 3.4 Hz, 4H), 7.51 (d, *J* = 8.5 Hz, 8H), 7.32 (d, *J* = 4.3 Hz, 4H), 7.05 (d, *J* = 8.9 Hz, 16H), 6.92 (d, *J* = 8.7 Hz, 8H), 6.85 (d, *J* = 9.0 Hz, 16H), 3.76 (s, 24H).

^{13}C NMR (125 MHz, THF-d8 [67.6 ppm], ppm) δ = 157.6 (8C), 149.7 (4C), 145.4 (4C), 144.0 (8C), 142.5 (4C), 141.7 (8C), 140.9 (4C), 135.3 (8C), 135.2

196

(4C), 132.9 (8C), 127.7 (16C), 127.4 (4C), 127.1 (8C), 125.8 (4C), 124.7 (8C), 123.9 (4C), 121.4 (8C), 119.8 (4C), 115.7 (16C), 55.8 (8C).

IR (ATR, $\tilde{v}$) = 3034, 2996, 2927, 2891, 2829, 1890, 1771, 1723, 1679, 1598, 1493, 1452, 1438, 1350, 1316, 1275, 1234, 1190, 1176, 1163, 1103, 1072, 1031, 1000, 970, 939, 911, 823, 792, 728, 711, 701, 574, 520 cm^{-1}.

ESI-MS (m/z): 2212 [M]$^+$, 1105 [M]$^{2+}$, 737 [M]$^{3+}$, 552 [M]$^{4+}$; HRMS–ESI (m/z): [M]$^+$ calcd for $C_{140}H_{104}N_8O_8S_4Ni_1$, 2212.62; found, 2212.63.

Meso-tetrakis-(4-(2-(3,4-ethylenedioxy)thienyl)-*N,N*-bis(4-methoxy-phenyl)aniline)porphyrin-Ni(II) (TM_45_Ni)

Meso-tetrakis-(4-(2-(3,4-ethylenedioxy)thienyl)-*N,N*-bis(4-methoxy-phenyl)aniline)porphyrin (TM_45_2H) (100 mg, 41.9 μmol, 1.00 equiv) and Ni(OAc)$_2$*4 H$_2$O (52.1 mg, 209 μmol, 5.00 equiv) was added to a 20 mL crimp cap vial and 12 mL of *N,N*-dimethylformamide were added. The reaction was heated to 140 °C for 22 h, and the reaction progress was monitored *via* TLC. The product mixture was cooled to 22 °C, and the solvent was removed under reduced

pressure. The crude product was purified *via* column chromatography on silica gel (eluent cyclohexane/tetrahydrofuran 3:1 → 0:1) and after evaporation of the solvent under reduced pressure washed with methanol to yield *meso*-tetrakis-(4-(2-(3,4-ethylenedioxy)thienyl)-*N,N*-bis(4-methoxyphenyl)aniline)porphyrin-Ni(II) (88.9 mg, 36.4 μmol, 87% yield) as a red solid.

M.p.: >300 °C.

R_f = 0.15 (cyclohexane/tetrahydrofuran).

^{1}H NMR (500 MHz, THF-d8 [3.58 ppm], ppm) δ = 8.83 (s, 8H), 8.11–8.10 (m, 8H), 8.00 (d, J = 8.2 Hz, 8H), 7.63 (d, J = 8.9 Hz, 8H), 7.04 (d, J = 9.0 Hz, 16H), 6.91 (d, J = 8.7 Hz, 8H), 6.85 (d, J = 9.0 Hz, 16H), 4.44 (dd, J = 27.8 Hz, J = 3.6 Hz, 16H), 3.76 (s, 24H).

^{13}C NMR (125 MHz, THF-d8 [67.6 ppm], ppm) δ = 157.4 (8C), 148.7 (4C), 144.0 (4C), 141.8 (8C), 141.0 (4C), 139.8 (4C), 139.2 (8C), 134.9 (8C), 134.2 (4C), 132.9 (8C), 127.7 (8C), 127.5 (16C), 126.4 (4C), 125.1 (8C), 121.4 (8C), 119.8 (4C), 117.1 (4C), 115.6 (16C), 114.1 (4C), 66.0 (4C), 65.8 (4C), 55.8 (8C).

IR (ATR, ṽ) = 2927, 2907, 2871, 2830, 1599, 1587, 1499, 1460, 1451, 1435, 1402, 1356, 1316, 1281, 1234, 1193, 1176, 1164, 1133, 1105, 1081, 1030, 1001, 967, 912, 881, 854, 822, 813, 796, 781, 722, 713, 697, 677, 592, 572, 564, 551, 516 cm^{-1}.

ESI-MS (m/z): 1221 [M]$^{2+}$, 814 [M]$^{3+}$, 610 [M]$^{4+}$; HRMS–ESI (m/z): [M]$^{2+}$ calcd for $C_{148}H_{112}N_8O_{16}S_4Ni_1$, 1221.82; found, 1221.83.

Meso-tetrakis-(9-(4-(2-thienyl)phenyl)-3,6-dimethoxycarbazole)porphyrin-Ni(II) (TM_46_Ni)

Meso-tetrakis-(9-(4-(2-thienyl)phenyl)-3,6-dimethoxycarbazole)porphyrin (TM_46_2H) (100 mg, 46.5 µmol, 1.00 equiv) and Ni(OAc)$_2$*4 H$_2$O (57.9 mg, 233 µmol, 5.00 equiv) were added to a 20 mL crimp cap vial and 12 mL of *N,N*-dimethylformamide was added. The reaction was heated to 140 °C for 22 h, and the reaction progress was monitored *via* TLC. The product mixture was cooled to 22 °C, and the solvent was removed under reduced pressure. The crude product was purified *via* column chromatography on silica gel (eluent cyclohexane/tetrahydrofuran 3:1 → 0:1) and after evaporation of the solvent under reduced pressure washed with methanol to yield *meso*-tetrakis-(9-(4-(2-thienyl)phenyl)-3,6-dimethoxycarbazole)porphyrin-Ni(II) (76.1 mg, 34.5 µmol, 74% yield) as a red solid.

M.p.: >300 °C.

R$_f$ = 0.15 (cyclohexane/tetrahydrofuran).

^{1}H NMR (500 MHz, THF-d8 [3.58 ppm], ppm) δ = 8.62 (s, 8H), 7.78 (s, 20H), 7.62 (s, 12H), 7.51 (s, 8H), 7.36–7.35 (m, 8H), 7.31 (d, *J* = 8.2 Hz, 8H), 6.95 (d, *J* = 7.8 Hz, 8H), 3.84 (s, 24H).

^{13}C NMR (125 MHz, THF-d8 [67.6 ppm], ppm) δ = 155.6 (8C), 144.2 (4C), 144.0 (4C), 143.9 (8C), 141.2 (4C), 138.7 (8C), 136.9 (8C), 135.4 (4C), 134.8 (8C), 133.6 (8C), 132.9 (8C), 127.7 (4C), 126.0 (4C), 125.8 (4C), 125.2 (8C), 124.7 (8C), 119.8 (8C), 116.0 (8C), 111.5 (8C), 103.8 (8C), 56.1 (8C).

IR (ATR, ṽ) = 2921, 2827, 1766, 1720, 1601, 1577, 1521, 1487, 1462, 1429, 1366, 1350, 1326, 1303, 1281, 1261, 1200, 1173, 1153, 1106, 1079, 1030, 1001, 972, 945, 911, 832, 790, 724, 711, 659, 584 cm^{-1}.

ESI-MS (m/z): 2204 [M+H]$^+$; HRMS–ESI (m/z): [M+H]$^+$ calcd for $C_{140}H_{97}N_8O_8S_4Ni_1$, 2204.57; found, 2204.57.

Meso-tetrakis-(9-(4-(2-(3,4-ethylenedioxy)thienyl)phenyl)-3,6-dimethoxy-carbazole)porphyrin-Ni(II) (TM_47_Ni)

Meso-tetrakis-(9-(4-(2-(3,4-ethylenedioxy)thienyl)phenyl)-3,6-dimethoxycarbazole)-porphyrin (TM_47_2H) (70.0 mg, 29.4 µmol, 1.00 equiv) and Ni(OAc)$_2$*4 H$_2$O (36.6 mg, 147 µmol, 5.00 equiv) were added to a 20 mL crimp cap vial and 7.61 mL of *N,N*-dimethylformamide were added. The reaction was heated to 140 °C for 16 h, and the reaction progress was monitored *via* TLC. The product mixture was cooled to 22 °C, and the solvent was removed under reduced pressure. The crude product was purified *via* column chromatography on silica gel (eluent cyclohexane/tetrahydrofuran 5:1 → 1:2) and after evaporation of the solvent under reduced pressure washed with methanol to yield *meso*-tetrakis-(9-(4-(2-(3,4-ethylenedioxy)thienyl)phenyl)-3,6-dimethoxycarbazole)porphyrin-Ni(II) (45.9 mg, 18.8 µmol, 64% yield) as a red solid.

M.p.: >300 °C.

R$_f$ = 0.2 (cyclohexane/tetrahydrofuran).

^{1}H NMR (500 MHz, THF-d8 [3.58 ppm], ppm) δ = 8.79 (s, 8H), 8.12 (d, J = 8.1 Hz, 8H), 8.05 (d, J = 8.4 Hz, 8H), 7.98 (d, J = 8.1 Hz, 8H), 7.64 (d, J = 2.3 Hz, 8H), 7.58 (d, J = 8.5 Hz, 8H), 7.37 (d, J = 8.9 Hz, 8H), 6.99 (dd, J = 2.4 Hz, J = 8.9 Hz, 8H), 4.48 (d, J = 2.1 Hz, 16H), 3.88 (s, 24H).

^{13}C NMR (125 MHz, THF-d8 [67.6 ppm], ppm) δ = 155.6 (8C), 143.9 (8C), 140.9 (8C), 140.7 (8C), 140.3 (4C), 137.7 (8C), 137.1 (8C), 135.1 (8C), 133.8 (4C), 132.9 (4C), 132.8 (4C), 128.2 (8C), 127.5 (8C), 125.3 (4C), 125.1 (8C), 119.9 (4C), 116.0 (8C), 115.8 (4C), 111.5 (8C), 103.8 (8C), 66.0 (8C), 56.2 (8C).

IR (ATR, ṽ) = 2932, 2827, 1717, 1598, 1581, 1506, 1487, 1460, 1429, 1402, 1356, 1326, 1306, 1285, 1261, 1200, 1173, 1152, 1135, 1108, 1081, 1028, 1001, 969, 946, 911, 882, 853, 832, 789, 754, 725, 713, 697, 690, 657, 596, 585 cm^{-1}.

ESI-MS (m/z): 2436 [M]$^+$, 1218 [M]$^{2+}$, 811 [M]$^{3+}$; HRMS–ESI (m/z): [M]$^+$ calcd for $C_{148}H_{104}N_8O_{16}S_4Ni_1$, 2436.59; found, 2436.58.

Meso-tetrakis-(4-(2-thienyl)-*N,N*-bis(4-methoxyphenyl)aniline)porphyrin-Co(II) (TM_44_Co)

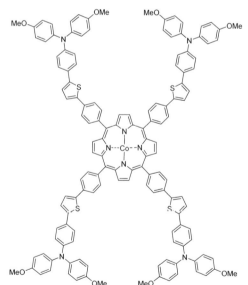

Meso-tetrakis-(4-(2-thienyl)-*N,N*-bis(4-methoxyphenyl)aniline)porphyrin (TM_44_2H) (100 mg, 46.4 µmol, 1.00 equiv) and Co(OAc)$_2$*4 H$_2$O (57.7 mg, 232 µmol, 5.00 equiv) was added to a 20 mL crimp cap vial and 12 mL of *N,N*-dimethylformamide was added. The reaction was heated to 140 °C for 22 h, and the reaction progress was monitored *via* TLC. The product mixture was cooled to 22 °C, and the solvent was removed under reduced pressure. The crude product was purified *via* column chromatography on silica gel (eluent cyclohexane/tetrahydrofuran 3:1 → 0:1)

and after evaporation of the solvent under reduced pressure washed with methanol to yield *meso*-tetrakis-(4-(2-thienyl)-*N*,*N*-bis(4-methoxy-phenyl)aniline)porphyrin-Co(II) (76.6 mg, 34.6 μmol, 75% yield) as a red solid.

M.p.: >300 °C.

R_f = 0.15 (cyclohexane/tetrahydrofuran).

^{1}H NMR (400 MHz, THF-d8 [3.58 ppm], ppm) δ = 8.88 (m, 8H), 8.11–8.01 (m, 8H), 7.72 (m, 8H), 7.60–7.53 (m, 8H), 7.14–7.06 (m, 24H), 6.91–6.85 (m, 24H), 3.80 (s, 24H).

^{13}C NMR (100 MHz, THF-d8 [67.6 ppm], ppm) δ = 157.6 (8C), 149.8 (8C), 145.8 (8C), 141.8 (8C), 127.8 (16C), 127.4 (8C), 121.6 (8C), 115.8 (16C), 55.9 (8C), due to paramagnetism of Co in porphyrin some carbon signals are missing (52C), structure defining carbons are visible.

IR (ATR, ṽ) = 2932, 2897, 2829, 2053, 2040, 1887, 1771, 1720, 1598, 1493, 1452, 1439, 1349, 1316, 1275, 1234, 1190, 1176, 1163, 1103, 1072, 1031, 999, 970, 939, 912, 823, 792, 728, 715, 700, 574, 493 cm^{-1}.

ESI-MS (*m/z*): 2213 [M]$^+$, 1106 [M]$^{2+}$, 737 [M]$^{3+}$, 553 [M]$^{4+}$; HRMS–ESI (*m/z*): [M]$^+$ calcd for $C_{140}H_{104}N_8O_8S_4Co_1$, 2213.62; found, 2213.63.

Meso-tetrakis-(4-(2-(3,4-ethylenedioxy)thienyl)-*N,N*-bis(4-methoxy-phenyl)aniline)porphyrin-Co(II) (TM_45_Co)

Meso-tetrakis-(4-(2-(3,4-ethylenedioxy)thienyl)-*N,N*-bis(4-methoxy-phenyl)aniline)porphyrin (TM_45_2H) (100 mg, 41.9 μmol, 1.00 equiv) and Co(OAc)$_2$*4 H$_2$O (52.1 mg, 209 μmol, 5.00 equiv) were added to a 20 mL crimp cap vial and 12 mL of *N,N*-dimethylformamide were added. The reaction was heated to 140 °C for 22 h, and the reaction progress was monitored *via* TLC. The product mixture was cooled to 22 °C, and the solvent was removed under reduced pressure. The crude product was purified *via* column chromatography on silica gel (eluent cyclohexane/tetrahydrofuran 3:1 → 0:1) and after evaporation of the solvent under reduced pressure washed with methanol to yield *meso*-tetrakis-(4-(2-(3,4-ethylenedioxy)thienyl)-*N,N*-bis(4-methoxyphenyl)aniline)porphyrin-Co(II) (80.0 mg, 32.7 μmol, 78% yield) as a reddish solid.

M.p.: >300 °C.

R_f = 0.1 (cyclohexane/tetrahydrofuran).

^{1}H NMR (400 MHz, THF-d8 [3.58 ppm], ppm) δ = 9.06 (s, 8H), 8.12–8.12 (m, 4H), 7.85–7.83 (m, 8H), 7.64 (m, 4H), 7.12 (m, 16H), 7.05–7.04 (m, 16H), 6.91–6.84 (m, 16H), 4.73–4.62 (m, 16H), 3.79 (s, 24H).

^{13}C NMR (100 MHz, THF-d8 [67.6 ppm], ppm) δ = 157.5 (8C), 148.8 (8C), 142.0 (8C), 141.5 (8C), 128.0 (8C), 127.6 (16C), 126.7 (8C), 121.6 (8C), 117.5 (8C), 115.7 (16C), 66.3 (4C), 66.1 (4C), 55.9 (8C), due to paramagnetism of Co in porphyrin some carbon signals are missing (36C), structure defining carbons are visible.

IR (ATR, $\tilde{v}$) = 2925, 2919, 2897, 2829, 1599, 1587, 1499, 1462, 1452, 1435, 1402, 1356, 1316, 1279, 1234, 1193, 1176, 1164, 1133, 1106, 1081, 1030, 999, 967, 912, 881, 854, 822, 813, 796, 781, 714, 677, 571, 551 cm^{-1}.

ESI-MS (m/z): 1222 [M]$^{2+}$, 814 [M]$^{3+}$, 611 [M]$^{4+}$; HRMS–ESI (m/z): [M]$^{2+}$ calcd for $C_{148}H_{112}N_8O_{16}S_4Co_1$, 1222.32; found, 1222.32.

Meso-tetrakis-(9-(4-(2-thienyl)phenyl)-3,6-dimethoxycarbazole)porphyrin-Co(II) (TM_46_Co)

Meso-tetrakis-(9-(4-(2-thienyl)phenyl)-3,6-dimethoxy-carbazole)porphyrin (TM_46_2H) (100 mg, 46.5 µmol, 1.00 equiv) and Co(OAc)$_2$*4 H$_2$O (58.0 mg, 233 µmol, 5.00 equiv) was added to a 20 mL crimp cap vial and 12 mL of *N,N*-dimethylformamide was added. The reaction was heated to 140 °C for 22 h, and the reaction progress was monitored *via* TLC. The product mixture was cooled to 22 °C, and the solvent was removed under reduced pressure. The crude product was purified *via* column chromatography on silica gel (eluent cyclohexane/tetrahydrofuran 3:1 → 0:1) and after evaporation of the solvent under reduced pressure washed with methanol to yield *meso*-tetrakis-(9-(4-(2-thienyl)phenyl)-3,6-dimethoxycarbazole)porphyrin-Co(II) (78.0 mg, 35.4 µmol, 76% yield) as a greenish solid.

M.p.: >300 °C.

R_f = 0.15 (cyclohexane/tetrahydrofuran).

^{1}H NMR (400 MHz, THF-d8 [3.58 ppm], ppm) δ = 8.16 (s, 8H), 7.97–7.93 (m, 10H), 7.77–7.66 (m, 30H), 7.50–7.48 (m, 8H), 7.42–7.39 (m, 8H), 7.08–7.01 (m, 8H), 3.93 (s, 24H).

^{13}C NMR (100 MHz, THF-d8 [67.6 ppm], ppm) δ = 155.7 (8C), 137.1 (8C), 128.1 (8C), 125.2 (8C), 116.3 (8C), 111.6 (8C), 103.8 (8C), 56.3 (8C), due to paramagnetism of Co in porphyrin some carbon signals are missing (76C), structure defining carbons are visible.

IR (ATR, ṽ) = 2927, 2827, 1717, 1604, 1579, 1513, 1487, 1462, 1429, 1366, 1349, 1326, 1303, 1281, 1259, 1200, 1174, 1153, 1105, 1069, 1028, 1000, 972, 945, 832, 790, 754, 722, 713, 659, 584 cm^{-1}.

ESI-MS (m/z): 2204 [M]$^+$, 1102 [M]$^{2+}$, 734 [M]$^{3+}$; HRMS–ESI (m/z): [M]$^+$ calcd for $C_{140}H_{96}N_8O_8S_4Co_1$, 2204.56; found, 2204.57.

Meso-tetrakis-(9-(4-(2-(3,4-ethylenedioxy)thienyl)phenyl)-3,6-dimethoxy-carbazole)porphyrin-Co(II) (TM_47_Co)

Meso-tetrakis-(9-(4-(2-(3,4-ethylenedioxy)thienyl)phenyl)-3,6-dimethoxycarbazole)-porphyrin (TM_47_2H) (70.0 mg, 29.4 µmol, 1.00 equiv) and Co(OAc)$_2$*4 H$_2$O (36.6 mg, 147 µmol, 5.00 equiv) were added to a 20 mL crimp cap vial and 7.61 mL of *N,N*-dimethylformamide were added. The reaction was heated to 140 °C for 16 h, and the reaction progress was monitored *via* TLC. The product mixture was cooled to 22 °C, and the solvent was removed under reduced pressure. The crude product was purified *via* column chromatography on silica gel (eluent cyclohexane/tetrahydrofuran 3:1 → 1:4) and after evaporation of the solvent under reduced pressure washed with methanol to yield *meso*-tetrakis-(9-(4-(2-(3,4-ethylenedioxy)thienyl)phenyl)-3,6-dimethoxycarbazole)porphyrin-Co(II) (35.0 mg, 14.4 µmol, 49% yield) as a red solid.

M.p.: >300 °C.

R$_f$ = 0.2 (cyclohexane/tetrahydrofuran).

^{1}H NMR (400 MHz, THF-d8 [3.58 ppm], ppm) δ = 8.95 (s, 8H), 8.30–8.21 (m, 8H), 8.03–7.97 (m, 8H), 7.75–7.59 (m, 8H), 7.50–7.48 (m, 8H), 7.41–7.36 (m, 8H), 7.09–7.07 (m, 8H), 7.03–6.99 (m, 8H), 4.81–4.45 (m, 16H), 3.93–3.89 (m, 24H).

^{13}C NMR (100 MHz, THF-d8 [67.6 ppm], ppm) δ = 155.6 (8C), 137.1 (8C), 131.4 (8C), 128.2 (8C), 127.4 (8C), 125.1 (8C), 116.1 (8C), 111.6 (8C), 104.0 (8C), 66.1 (4C), 65.9 (4C), 56.2 (8C), due to paramagnetism of Co in porphyrin some carbon signals are missing (60C), structure defining carbons are visible.

IR (ATR, ṽ) = 2924, 2836, 1718, 1507, 1487, 1462, 1432, 1405, 1357, 1327, 1307, 1285, 1262, 1201, 1173, 1153, 1135, 1108, 1079, 1028, 1001, 969, 946, 911, 878, 851, 833, 788, 727, 717, 697, 657 cm^{-1}.

ESI-MS (m/z): 2436 [M]$^+$; HRMS–ESI (m/z): [M]$^+$ calcd for C$_{148}$H$_{104}$N$_8$O$_{16}$S$_4$Co$_1$, 2436.58; found, 2436.58.

5.2.5 Tetra-*para*-substituted Tetraphenylethenes

1,1,2,2-Tetrakis(4-bromophenyl)ethene (49)

Bromine (9.61 g, 3.09 mL, 60.2 mmol, 8.00 equiv) was added to a solution of 1,1,2,2-tetrakis(phenyl)ethene (2.50 g, 7.52 mmol, 1.00 equiv) in a mixture of acetic acid/di-chloromethane (7 mL/15 mL) at 0 °C. The reaction mixture was stirred at 22 °C for 4 h. The product mixture was poured into water and extracted with dichloromethane (3x). The combined organic phases were washed with sat. NaHSO$_3$ solution, water, and brine and dried over anhydrous Na$_2$SO$_4$. The solvent was removed under reduced pressure, and the crude product was recrystallized from methanol to yield 1,1,2,2-tetrakis(4-bromophenyl)ethene (4.26 g, 6.57 mmol, 87% yield) as a colorless solid.

M.p.: 254 °C.

R$_f$ = 0.9 (cyclohexane/ethyl acetate 4:1).

^{1}H NMR (500 MHz, Chloroform-d [7.27 ppm], ppm) δ = 7.28–7.27 (m, 8H), 6.85 (d, J = 8.5 Hz, 8H).

^{13}C NMR (125 MHz, Chloroform-d [77.0 ppm], ppm) δ = 141.5 (4C), 139.6 (2C), 132.7 (8C), 131.3 (8C), 121.3 (4C).

IR (ATR, ṽ) = 1904, 1579, 1483, 1392, 1102, 1069, 1007, 861, 817, 792, 735, 717, 632 cm^{-1}.

FAB-MS (m/z, 3-NBA): 650/648/646 (1/1/1) [M]$^+$, 195 (15), 155/154 (30/100), 138/137/136 (34/66/66); HRMS–FAB (m/z): [M]$^+$ calcd for $C_{26}H_{18}{}^{79}Br_4$, 645.8136, found 645.8139.

4-Pinacolboryl-*N*,*N*-bis(4-methoxyphenyl)aniline (50)

4-Bromo-*N*,*N*-bis(4-methoxyphenyl)aniline (1) (800 mg, 2.08 mmol, 1.00 equiv), Pd(dppf)Cl$_2$ (152 mg, 208 μmol, 0.100 equiv), KOAc (613 mg, 6.25 mmol, 3.00 equiv) and bis(pinacolato)diboron (793 mg, 3.12 mmol, 1.50 equiv) were added to a crimp cap vial and refilled with argon three times. Dry 1,4-dioxane (13 mL) was added, and the reaction mixture was stirred at 100 °C for 18 h. The product mixture was cooled to 22 °C, poured into water, and extracted with dichloromethane. The combined organic phases were washed with water (3x) and brine (3x), dried over anhydrous Na$_2$SO$_4$, and the solvent was removed under reduced pressure. The crude product was purified *via* column chromatography on silica gel (eluent cyclohexane/ethyl acetate 10:1 → 7:1) to yield 4-pinacolboryl-*N*,*N*-bis(4-methoxyphenyl)aniline (487 mg, 1.13 mmol, 54% yield) as a brown solid.

M.p.: 128 °C.

R$_f$ = 0.5 (cyclohexane/ethyl acetate 4:1).

^{1}H NMR (500 MHz, Chloroform-d [7.27 ppm], ppm) δ = 7.61 (d, J = 8.4 Hz, 2H), 7.07 (d, J = 8.7 Hz, 4H), 6.88–6.83 (m, 6H), 3.81 (s, 6H), 1.33 (s, 12H).

^{13}C NMR (125 MHz, Chloroform-d [77.0 ppm], ppm) δ = 156.2 (2C), 151.3 (1C), 140.4 (2C), 135.7 (2C), 127.1 (4C), 118.6 (2C), 114.7 (4C), 83.4 (2C), 55.5 (2C), 24.8 (4C), C-B was missing due to interactions (1C).

IR (ATR, $\tilde{v}$) = 3055, 2997, 2979, 2958, 2932, 2910, 2836, 1609, 1596, 1557, 1503, 1462, 1439, 1421, 1394, 1360, 1327, 1315, 1285, 1242, 1232, 1197, 1179, 1166, 1142, 1106, 1092, 1031, 1011, 960, 915, 860, 841, 830, 816, 798, 779, 734, 717, 684, 676, 657, 646, 595, 578, 547 cm^{-1}.

FAB-MS (m/z, 3-NBA): 432/431/430 (39/100/25) [M]$^+$; HRMS–FAB (m/z): [M]$^+$ calcd for $C_{26}H_{30}O_4N_1B_1$, 431.2262, found 431.2261.

4-Pinacolboryl-*N,N*-bis(3-methoxyphenyl)aniline (51)

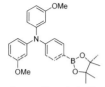

4-Bromo-*N,N*-bis(3-methoxyphenyl)aniline (2) (750 mg, 1.95 mmol, 1.00 equiv), Pd(dppf)Cl$_2$ (143 mg, 195 μmol, 0.100 equiv), KOAc (575 mg, 5.86 mmol, 3.00 equiv) and bis(pinacolato)diboron (743 mg, 2.93 mmol, 1.50 equiv) were added to a crimp cap vial and refilled with argon three times. Dry 1,4-dioxane (13 mL) was added, and the reaction mixture was stirred at 100 °C for 18 h. The product mixture was cooled to 22 °C, poured into water, and extracted with dichloromethane. The combined organic phases were washed with water (3x) and brine (3x), dried over anhydrous Na$_2$SO$_4$, and the solvent was removed under reduced pressure. The crude product was purified *via* column chromatography on silica gel (eluent cyclohexane/ethyl acetate 10:1) to yield 4-pinacolboryl-*N,N*-bis(3-methoxyphenyl)aniline (751 mg, 1.74 mmol, 89% yield) as a brown solid.

M.p.: 79 °C.

R_f = 0.6 (cyclohexane/ethyl acetate 4:1).

^{1}H NMR (500 MHz, Chloroform-d [7.27 ppm], ppm) δ = 7.67 (d, J = 8.4 Hz, 2H), 7.16 (t, J = 8.1 Hz, 2H), 7.06 (d, J = 8.4 Hz, 2H), 6.70–6.66 (m, 4H), 6.61 (dd, J = 2.1 Hz, J = 8.2 Hz, 2H), 3.72 (s, 6H), 1.34 (s, 12H).

^{13}C NMR (125 MHz, Chloroform-d [77.0 ppm], ppm) δ = 160.4 (2C), 150.3 (1C), 148.5 (2C), 135.8 (2C), 129.8 (2C), 122.2 (2C), 117.4 (2C), 110.6 (2C), 109.0 (2C), 83.6 (2C), 55.2 (2C), 24.9 (4C), C-B was missing due to interactions (1C).

IR (ATR, $\tilde{v}$) = 2975, 2935, 2833, 1584, 1557, 1511, 1487, 1463, 1451, 1432, 1395, 1357, 1315, 1289, 1275, 1248, 1210, 1174, 1154, 1140, 1106, 1089, 1038, 1013, 990, 960, 857, 834, 783, 773, 737, 701, 690, 673, 656, 603, 592, 578, 555, 520 cm^{-1}.

FAB-MS (*m/z*, 3-NBA): 432/431/430 (53/100/25) [M]$^+$; HRMS–FAB (*m/z*): [M]$^+$ calcd for C$_{26}$H$_{30}$O$_4$N$_1$B$_1$, 431.2262, found 431.2261.

4-Pinacolboryl-*N,N*-bis(2-methoxyphenyl)aniline (52)

4-Bromo-*N,N*-bis(2-methoxyphenyl)aniline (3) (200 mg, 520 μmol, 1.00 equiv), Pd(dppf)Cl$_2$ (38.1 mg, 52.0 μmol, 0.100 equiv), KOAc (153 mg, 1.56 mmol, 3.00 equiv) and bis(pinacolato)diboron (198 mg, 781 μmol, 1.50 equiv) were added to a crimp cap vial and refilled with argon three times. Dry 1,4-dioxane (15 mL) was added, and the reaction mixture was stirred at 100 °C for 18 h. The product mixture was cooled to 22 °C, poured into water, and extracted with dichloromethane. The combined organic phases were washed with water (3x) and brine (3x), dried over anhydrous Na$_2$SO$_4$, and the solvent was removed under reduced pressure. The crude product was purified *via* column chromatography on silica gel (eluent cyclohexane/ethyl acetate 10:1 → 5:1) to yield 4-pinacolboryl-*N,N*-bis(2-methoxyphenyl)aniline (195 mg, 452 μmol, 87% yield) as a brown solid.

M.p.: 184 °C.

R_f = 0.5 (cyclohexane/ethyl acetate 4:1).

^{1}H NMR (500 MHz, Chloroform-d [7.27 ppm], ppm) δ = 7.56 (d, *J* = 8.7 Hz, 2H), 7.24–7.17 (m, 4H), 6.96–6.89 (m, 4H), 6.56 (d, *J* = 8.5 Hz, 2H), 3.69 (s, 6H), 1.31 (s, 12H).

^{13}C NMR (125 MHz, Chloroform-d [77.0 ppm], ppm) δ = 155.7 (2C), 150.8 (1C), 135.3 (2C), 134.8 (2C), 129.3 (2C), 126.6 (2C), 121.2 (2C), 115.3 (2C), 112.7 (2C), 83.1 (2C), 55.7 (2C), 24.8 (4C), C-B was missing due to interaction (1C).

IR (ATR, ṽ) = 2973, 2932, 2839, 1605, 1589, 1496, 1460, 1391, 1360, 1332, 1323, 1283, 1272, 1238, 1210, 1174, 1143, 1119, 1089, 1044, 1027, 959, 860, 849, 824, 755, 735, 652, 635, 575, 545 cm^{-1}.

FAB-MS (*m/z*, 3-NBA): 431 (100) [M]$^+$, 325 (55), 95 (59); HRMS–FAB (*m/z*): [M]$^+$ calcd for C$_{26}$H$_{30}$O$_4$N$_1$B$_1$, 431.2262, found 431.2263.

4-Pinacolboryl-*N,N*-bis(2,4-dimethoxyphenyl)aniline (53)

4-Bromo-*N,N*-bis(2,4-dimethoxyphenyl)aniline (4) (800 mg, 1.80 mmol, 1.00 equiv), Pd(dppf)Cl$_2$ (132 mg, 180 μmol, 0.100 equiv), KOAc (530 mg, 5.40 mmol, 3.00 cquiv) and bis(pinacolato)diboron (686 mg, 2.70 mmol, 1.50 equiv) were added to a crimp cap vial and refilled with argon three times. Dry 1,4-dioxane (12 mL) was added, and the reaction mixture was stirred at 100 °C for 18 h. The product mixture was cooled to 22 °C, poured into water, and extracted with dichloromethane. The combined organic phases were washed with water (3x) and brine (3x), dried over anhydrous Na$_2$SO$_4$, and the solvent was removed under reduced pressure. The crude product was purified *via* column chromatography on silica gel (eluent cyclohexane/ethyl acetate 10:1 → 7:1) to yield 4-pinacolboryl-*N,N*-bis(2,4-dimethoxyphenyl)aniline (576 mg, 1.17 mmol, 65% yield) as a brown solid.

M.p.: 188 °C.

R$_f$ = 0.2 (cyclohexane/ethyl acetate 4:1).

^{1}H NMR (500 MHz, Chloroform-d [7.27 ppm], ppm) δ = 7.54 (d, J = 8.5 Hz, 2H), 7.27–7.26 (m, 2H), 6.53 (d, J = 2.6 Hz, 2H), 6.47–6.42 (m, 4H), 3.82 (s, 6H), 3.72 (s, 6H), 1.31 (s, 12H).

^{13}C NMR (125 MHz, Chloroform-d [77.0 ppm], ppm) δ = 158.9 (2C), 157.0 (2C), 151.7 (1C), 135.5 (2C), 130.6 (2C), 127.6 (2C), 113.3 (2C), 104.7 (2C), 99.9 (2C), 83.0 (2C), 55.7 (2C), 55.4 (2C), 24.8 (4C), C-B was missing due to interaction (1C).

IR (ATR, ṽ) = 2993, 2973, 2928, 2874, 2846, 2837, 1599, 1582, 1551, 1506, 1480, 1460, 1439, 1417, 1391, 1358, 1341, 1313, 1300, 1275, 1254, 1239, 1208, 1183, 1159, 1143, 1132, 1089, 1037, 1007, 960, 949, 933, 905, 882, 858, 844, 834, 817, 806, 741, 734, 718, 684, 670, 653, 637, 606, 588, 564, 537, 518 cm^{-1}.

FAB-MS (*m/z*, 3-NBA): 492/491/490 (41/100/25) [M]$^+$; HRMS–FAB (*m/z*): [M]$^+$ calcd for C$_{28}$H$_{34}$O$_6$N$_1$B$_1$, 491.2479, found 491.2462.

4-Pinacolboryl-*N,N*-bis(3,4,5-trimethoxyphenyl)aniline (54)

4-Bromo-*N,N*-bis(3,4,5-trimethoxyphenyl)aniline (5) (800 mg, 1.59 mmol, 1.00 equiv), Pd(dppf)Cl$_2$ (116 mg, 159 μmol, 0.100 equiv), KOAc (467 mg, 4.76 mmol, 3.00 equiv) and di(pinacolato)diboron (604 mg, 2.38 mmol, 1.50 equiv) were added to a crimp cap vial and refilled with argon three times. Dry 1,4-dioxane (12 mL) was added, and the reaction mixture was stirred at 100 °C for 18 h. The product mixture was cooled to 22 °C, poured into water, and extracted with dichloromethane. The combined organic phases were washed with water (3x) and brine (3x), dried over anhydrous Na$_2$SO$_4$, and the solvent was removed under reduced pressure. The crude product was purified *via* column chromatography on silica gel (eluent cyclohexane/ethyl acetate 10:1 → 2:1) to yield 4-pinacolboryl-*N,N*-bis(3,4,5-trimethoxyphenyl)aniline (847 mg, 1.54 mmol, 97% yield) as a brown solid.

M.p.: 142 °C.

R$_f$ = 0.1 (cyclohexane/ethyl acetate 4:1).

^{1}H NMR (500 MHz, Chloroform-d [7.27 ppm], ppm) δ = 7.66 (d, *J* = 8.5 Hz, 2H), 7.02 (d, *J* = 8.5 Hz, 2H), 6.33 (s, 4H), 3.86 (s, 6H), 3.72 (s, 12H), 1.35 (s, 12H).

^{13}C NMR (125 MHz, Chloroform-d [77.0 ppm], ppm) δ = 153.6 (4C), 150.4 (1C), 143.0 (2C), 135.7 (2C), 134.5 (2C), 121.1 (2C), 102.7 (4C), 83.6 (2C), 61.0 (2C), 56.1 (4C), 24.8 (4C), C-B was missing due to interactions (1C).

IR (ATR, ṽ) = 2978, 2962, 2927, 1619, 1585, 1554, 1500, 1462, 1451, 1432, 1422, 1408, 1390, 1357, 1316, 1279, 1228, 1210, 1181, 1123, 1089, 1068, 1051, 1021, 1003, 989, 962, 922, 858, 824, 806, 754, 738, 698, 670, 654, 619 cm^{-1}.

FAB-MS (*m/z*, 3-NBA): 552/551 (49/100) [M]$^+$; HRMS–FAB (*m/z*): [M]$^+$ calcd for C$_{30}$H$_{38}$O$_8$N$_1$B$_1$, 551.2685, found 551.2683.

1,1,2,2-Tetrakis(4-phenyl-*N*,*N*-bis(4-methoxyphenyl)aniline)ethene (55)

1,1,2,2-Tetrakis(4-bromo-phenyl)ethene (49) (150 mg, 231 μmol, 1.00 equiv), Pd(dppf)Cl$_2$ (33.9 mg, 46.3 μmol, 0.200 equiv), K$_2$CO$_3$ (384 mg, 2.78 mmol, 12.0 equiv) and 4-pinacolboryl-*N*,*N*-bis(4-methoxyphenyl)aniline (50) (479 mg, 1.11 mmol, 4.80 equiv) were added to a crimp cap vial and refilled with argon three times. Degassed dimethyl sulfoxide (15 mL) was added, and the reaction mixture was stirred at 90 °C for 18 h. The product mixture was cooled to 22 °C and poured into water. The mixture was extracted with dichloromethane (3x), and the combined organic phases were washed with water (3x) and brine (3x) and dried over anhydrous Na$_2$SO$_4$. The solvent was removed under reduced pressure, and the crude product was purified *via* column chromatography on silica gel (eluent cyclohexane/dichloromethane/ethyl acetate 10:1:0 → 0:4:1) to yield 1,1,2,2-tetrakis(4-phenyl-*N*,*N*-bis(4-methoxy-phenyl)aniline)ethene (132 mg, 85.4 μmol, 37% yield) as a bright yellow solid.

M.p.: 139 °C.

R$_f$ = 0.7 (cyclohexane/ethyl acetate 1:1).

^{1}H NMR (500 MHz, Chloroform-d [7.27 ppm], ppm) δ = 7.40 (d, *J* = 8.7 Hz, 8H), 7.33 (d, *J* = 8.2 Hz, 8H), 7.13 (d, *J* = 8.2 Hz, 8H), 7.08 (d, *J* = 8.7 Hz, 16H), 6.97 (d, *J* = 8.5 Hz, 8H), 6.84 (d, *J* = 8.9 Hz, 16H), 3.81 (s, 24H).

^{13}C NMR (125 MHz, Chloroform-d [77.0 ppm], ppm) δ = 155.7 (8C), 147.9 (4C), 142.2 (4C), 140.9 (8C), 140.2 (2C), 138.3 (4C), 132.5 (4C), 131.9 (8C), 127.2 (8C), 126.4 (16C), 125.5 (8C), 120.9 (8C), 114.6 (16C), 55.4 (8C).

IR (ATR, ṽ) = 3026, 2946, 2928, 2902, 2832, 1599, 1500, 1490, 1460, 1439, 1397, 1360, 1317, 1276, 1235, 1197, 1177, 1164, 1143, 1105, 1033, 1004, 979, 962, 912, 820, 806, 781, 768, 752, 717, 698, 656, 598, 575 cm^{-1}.

ESI-MS (m/z): 1546/1545 (13/11) [M]$^+$, 1403 (17), 1367 (31), 1319 (100); HRMS–ESI (m/z): [M]$^+$ calcd for $C_{106}H_{84}N_4O_8$, 1544.6602; found, 1544.6585.

1,1,2,2-Tetrakis(4-phenyl-N,N-bis(3-methoxyphenyl)aniline)ethene (56)

1,1,2,2-Tetrakis(4-bromo-phenyl)ethene (49) (100 mg, 154 µmol, 1.00 equiv), Pd(dppf)Cl$_2$ (22.6 mg, 30.9 µmol, 0.200 equiv), K$_2$CO$_3$ (256 mg, 1.85 mmol, 12.0 equiv) and 4-pinacolboryl-N,N-bis(3-methoxy-phenyl)aniline (51) (319 mg, 741 µmol, 4.80 equiv) were added to a crimp cap vial and refilled with argon three times. Degassed dimethyl sulfoxide (15 mL) was added, and the reaction mixture was stirred at 90 °C for 18 h. The product mixture was cooled to 22 °C and poured into water. The mixture was extracted with dichloromethane (3x), and the combined organic phases were washed with water (3x) and brine (3x) and dried over anhydrous Na$_2$SO$_4$. The solvent was removed under reduced pressure, and the crude product was purified *via* column chromatography on silica gel (eluent cyclo-hexane/dichloromethane/ethyl acetate 10:1:0 → 0:4:1) to yield 1,1,2,2-tetrakis(4-phenyl-N,N-bis(3-methoxyphenyl)aniline)ethene (149 mg, 96.4 µmol, 62% yield) as a bright yellow solid.

M.p.: 124 °C.

R$_f$ = 0.8 (cyclohexane/ethyl acetate 1:1).

^{1}H NMR (500 MHz, Chloroform-d [7.27 ppm], ppm) δ = 7.48 (d, J = 8.5 Hz, 8H), 7.38 (d, J = 8.2 Hz, 8H), 7.18–7.12 (m, 24H), 6.71–6.67 (m, 16H), 6.60–6.58 (m, 8H), 3.73 (s, 24H).

^{13}C NMR (125 MHz, Chloroform-d [77.0 ppm], ppm) δ = 160.4 (8C), 148.7 (8C), 146.8 (4C), 142.5 (4C), 140.3 (2C), 138.3 (4C), 134.8 (4C), 132.0 (8C), 129.8 (8C), 127.4 (8C), 125.8 (8C), 124.4 (8C), 116.8 (8C), 110.1 (8C), 108.4 (8C), 55.2 (8C).

IR (ATR, ṽ) = 2953, 2931, 2917, 2904, 2832, 1735, 1588, 1521, 1485, 1466, 1449, 1436, 1383, 1361, 1307, 1273, 1241, 1205, 1173, 1159, 1137, 1082, 1040, 1001, 996, 980, 960, 822, 806, 764, 735, 690, 646, 630, 605, 578, 565, 550 cm^{-1}.

ESI-MS (*m/z*): 1548/1547/1546/1545 (20/58/100/83) [M]$^+$; HRMS–ESI (*m/z*): [M]$^+$ calcd for C$_{106}$H$_{84}$N$_4$O$_8$, 1544.6602; found, 1544.6585.

1,1,2,2-Tetrakis(4-phenyl-*N,N*-bis(2-methoxyphenyl)aniline)ethene (57)

1,1,2,2-Tetrakis(4-bromophenyl)ethene (49) (100 mg, 154 μmol, 1.00 equiv), Pd(dppf)Cl$_2$ (22.6 mg, 30.9 μmol, 0.200 equiv), K$_2$CO$_3$ (256 mg, 1.85 mmol, 12.0 equiv) and 4-pinacolboryl-*N,N*-bis(2-methoxyphenyl)aniline (52) (293 mg, 679 μmol, 4.40 equiv) were added to a crimp cap vial and refilled with argon three times. Degassed dimethyl sulfoxide (15 mL) was added, and the reaction mixture was stirred at 90 °C for 18 h. The product mixture was cooled to 22 °C and poured into water. The mixture was extracted with dichloromethane (3x), and the combined organic phases were washed with water (3x) and brine (3x) and dried over anhydrous Na$_2$SO$_4$. The solvent was removed under reduced pressure, and the crude product was purified *via* column chromatography on silica gel (eluent cyclohexane/dichloromethane/ethyl acetate 10:1:0 → 0:4:1) to yield 1,1,2,2-tetrakis(4-phenyl-*N,N*-bis(2-methoxyphenyl)aniline)ethene (169 mg, 109 μmol, 71% yield) as a bright yellow solid.

M.p.: 166 °C.

R$_f$ = 0.7 (cyclohexane/ethyl acetate 1:1).

^{1}H NMR (500 MHz, Chloroform-d [7.27 ppm], ppm) δ = 7.34 (dd, J = 13.1 Hz, J = 8.5 Hz, 16H), 7.20–7.15 (m, 16H), 7.10 (d, J = 8.2 Hz, 8H), 6.96–6.90 (m, 16H), 6.67 (d, J = 8.5 Hz, 8H), 3.68 (s, 24H).

^{13}C NMR (125 MHz, Chloroform-d [77.0 ppm], ppm) δ = 155.3 (8C), 147.6 (4C), 142.0 (4C), 140.0 (2C), 138.6 (4C), 135.5 (8C), 131.9 (8C), 131.2 (8C), 128.7 (8C), 126.6 (8C), 126.0 (8C), 125.3 (4C), 121.2 (8C), 117.5 (8C), 112.8 (8C), 55.8 (8C).

IR (ATR, $\tilde{v}$) = 3026, 2932, 2832, 1731, 1723, 1608, 1589, 1520, 1490, 1453, 1435, 1402, 1319, 1290, 1266, 1232, 1196, 1179, 1159, 1112, 1045, 1023, 1003, 977, 817, 805, 744, 696, 619, 605, 568 cm^{-1}.

ESI-MS (m/z): 1546 (16) [M]$^+$, 864 (36), 818 (93), 773 (100), 608 (70), 522 (64), 489 (21), 431 (46); HRMS–ESI (m/z): [M]$^+$ calcd for $C_{106}H_{84}N_4O_8$, 1544.6602; found, 1544.6585.

1,1,2,2-Tetrakis(4-phenyl-N,N-bis(2,4-dimethoxyphenyl)aniline)ethene (58)

1,1,2,2-Tetrakis(4-bromo-phenyl)ethene (49) (100 mg, 154 μmol, 1.00 equiv), Pd(dppf)Cl$_2$ (22.6 mg, 30.9 μmol, 0.200 equiv), K$_2$CO$_3$ (256 mg, 1.85 mmol, 12.0 equiv) and 4-pinacolboryl-N,N-bis(2,4-dimethoxy-phenyl)aniline (53) (364 mg, 741 μmol, 4.80 equiv) were added to a crimp cap vial and refilled with argon three times. Degassed dimethyl sulfoxide (15 mL) was added, and the reaction mixture was stirred at 90 °C for 18 h. The product mixture was cooled to 22 °C and poured into water. The mixture was extracted with dichloromethane (3x), and the combined organic phases were washed with water (3x) and brine (3x) and dried over anhydrous Na$_2$SO$_4$. The solvent was removed under reduced pressure, and the crude product was purified *via* column chromatography on silica gel (eluent cyclohexane/ethyl acetate 10:1 → 1:1) to yield 1,1,2,2-tetrakis(4-phenyl-*N,N*-bis(2,4-dimethoxyphenyl)aniline)ethene (149 mg, 96.4 µmol, 62% yield) as a bright yellow solid.

M.p.: 148 °C.

R_f = 0.3 (cyclohexane/ethyl acetate 1:1).

^{1}H NMR (500 MHz, Chloroform-d [7.27 ppm], ppm) δ = 7.30 (dd, J = 12.1 Hz, J = 8.7 Hz, 16H), 7.23 (d, J = 8.5 Hz, 8H), 7.06 (d, J = 8.2 Hz, 8H), 6.53–6.50 (m, 16H), 6.46 (dd, J = 2.6 Hz, J = 8.5 Hz, 8H), 3.82 (s, 24H), 3.72 (s, 24H).

^{13}C NMR (125 MHz, Chloroform-d [77.0 ppm], ppm) δ = 158.5 (8C), 156.8 (8C), 148.4 (4C), 141.8 (4C), 139.9 (2C), 138.7 (4C), 131.8 (8C), 130.2 (8C), 129.7 (4C), 128.2 (8C), 126.7 (8C), 125.1 (8C), 115.0 (8C), 104.7 (8C), 100.1 (8C), 55.7 (8C), 55.4 (8C).

IR (ATR, ṽ) = 2929, 2833, 1602, 1582, 1504, 1492, 1462, 1453, 1436, 1415, 1333, 1292, 1275, 1238, 1204, 1156, 1120, 1030, 941, 817, 802, 755, 731, 720, 633, 565 cm^{-1}.

ESI-MS (*m/z*): 1787/1786/1785 (62/100/75) [M]$^+$, 1559 (51); HRMS–ESI (*m/z*): [M]$^+$ calcd for $C_{114}H_{104}N_4O_{16}$, 1784.7447; found, 1784.7455.

1,1,2,2-Tetrakis(4-phenyl-*N,N*-bis(3,4,5-trimethoxyphenyl)aniline)ethene (59)

1,1,2,2-Tetrakis(4-bromophenyl)ethene (49) (150 mg, 231 μmol, 1.00 equiv), Pd(dppf)Cl$_2$ (33.9 mg, 46.3 μmol, 0.200 equiv), K$_2$CO$_3$ (384 mg, 2.78 mmol, 12.0 equiv) and 4-pinacolboryl-*N,N*-bis(3,4,5-trimethoxyphenyl)aniline (55) (613 mg, 1.11 mmol, 4.80 equiv) were added to a crimp cap vial and refilled with argon three times. Degassed dimethyl sulfoxide (15 mL) was added, and the reaction mixture was stirred at 90 °C for 18 h. The product mixture was cooled to 22 °C and poured into water. The mixture was extracted with dichloromethane (3x), and the combined organic phases were washed with water (3x) and brine (3x) and dried over anhydrous Na$_2$SO$_4$. The solvent was removed under reduced pressure, and the crude product was purified *via* column chromatography on silica gel (eluent cyclohexane/ethyl acetate 10:1 → 1:1) to yield 1,1,2,2-tetrakis(4-phenyl-*N,N*-bis(3,4,5-trimethoxyphenyl)aniline)ethene (206 mg, 102 μmol, 44% yield) as a bright yellow solid.

M.p.: 165 °C.

R$_f$ = 0.1 (cyclohexane/ethyl acetate 1:1).

^{1}H NMR (500 MHz, Chloroform-d [7.27 ppm], ppm) δ = 7.47 (d, *J* = 8.7 Hz, 8H), 7.40 (d, *J* = 8.2 Hz, 8H), 7.17 (d, *J* = 8.2 Hz, 8H), 7.09 (d, *J* = 8.7 Hz, 8H), 6.33 (s, 16H), 3.86 (s, 24H), 3.72 (s, 48H).

^{13}C NMR (125 MHz, Chloroform-d [77.0 ppm], ppm) δ = 153.6 (16C), 147.1 (4C), 143.3 (8C), 142.5 (4C), 140.2 (2C), 138.2 (4C), 134.2 (8C), 133.8 (4C),

132.0 (8C), 127.2 (8C), 125.6 (8C), 123.2 (8C), 102.3 (16C), 61.0 (8C), 56.2 (16C).

IR (ATR, $\tilde{v}$) = 2931, 2833, 2826, 1737, 1584, 1490, 1460, 1446, 1426, 1409, 1363, 1347, 1265, 1227, 1207, 1180, 1122, 1102, 1050, 1001, 924, 856, 819, 805, 771, 754, 727, 696, 660, 649, 628 cm^{-1}.

ESI-MS (*m/z*): 2027/2026/2025 (67/100/61) [M]$^+$; HRMS–ESI (*m/z*): [M]$^+$ calcd for $C_{122}H_{120}N_4O_{24}$, 2024.8293; found, 2024.8316.

6 ABBREVIATIONS

3-NBA	3-Nitrobenzyl alcohol
A	Ampere
Ac	Acetyl
aq.	Aqueous
ATR	Attenuated total reflection
BP	British Petroleum
CB	Conduction band
Cbz	Carbazole
CIGS	Copper indium gallium diselenide
conc.	Concentrated
CPL	Circularly polarized luminescence
CV	Cyclic voltammetry
CVD	Chemical vapor deposition
Cy	Cyclohexyl
d	Days
DFT	Density functional theory
DMF	N,N-Dimethylformamide
DMSO	Dimethyl sulfoxide
dppf	1,1'-Bis(diphenylphosphino)ferrocene
DSSC	Dye-sensitized solar cell
EDOT	3,4-Ethylenedioxythiophene
EI	Electron impact
EJ	Exajoules
equiv.	Equivalents
ESI	Electron spray ionization

ETL	Electron transport layer
ETM	Electron transport material
eV	Electronvolt
F4-TCNQ	2,3,5,6-Tetrafluoro-7,7,8,8-tetracyanoquindodimethane
FA	Formamidinium
FAB	Fast atom bombardement
Fc	Ferrocene
FDT	Fluorene-dithiophene
FF	Fill Factor
GC	Glassy carbon
h	Hours
HOMO	Highest occupied molecular orbital
HRMS	High-resolution mass spectrometry
HSQC	Heteronuclear single-quantum coherence
HTL	Hole transport layer
HTM	Hole transport material
Hz	Hertz
ICT	Intramolecular charge transfer
IP	Ionization potential
iPr	*Iso*-propyl
IR	Infrared
ITO	Indium tin oxide
IUPAC	International Union of Pure and Applied Chemistry
LUMO	Lowest unoccupied molecular orbital
M	Molar
m	*Meta*
m.p.	Melting point
MA	Methylammonium

MAPI	Methylammonium lead iodide
min	Minute
mL	Milliliter
mmol	Millimole
MPP	Maximum power point
MS	Mass spectrometry
n-BuLi	*n*-Butyllithium
NMR	Nuclear magnetic resonance
o	*Ortho*
OLED	Organic light-emitting diode
OMe	Methoxy
OSC	Organic solar cell
p	*Para*
P3CT	Poly[3-(4-carboxybutyl)thiophene-2,5-diyl]
P3HT	Poly(3-hexylthiophene)
PCE	Power conversion efficiency
PCP	[2.2]paracyclophane
PEDOT	Poly(3,4-ethylenedioxythiophene)
PEPPSI	Pyridine-enhanced precatalyst preparation stabilization and initiation
PESA	Photoemission spectroscopy in air
Ph	Phenyl
PivOH	Pivalic acid
PL	Photoluminescence
ppm	Parts per million
PSC	Perovskite solar cell
PSS	Poly(styrenesulfonate)
PTAA	Poly(bis(4-phenyl)(2,4,6-trimethylphenyl)amine

R_f	Retardation factor
SOHEL	Self-organized hole extraction layer
TADF	Thermally activated delayed fluorescence
TBP	*tert*-Butylpyridine
TFA	Trifluoroacetic acid
TFSI	Bis(trifluoromethanesulfonyl)imide
TGA	Thermogravimetric analysis
THF	Tetrahydrofuran
TIPS	Triisopropylsilyl
TLC	Thin layer chromatography
TM	Target material
Tol	Tolyl
TPA	Triphenylamine
TPD	*N,N'*-Bis(4-butylphenyl)-*N,N'*-bis(phenyl)benzidine
TPE	Tetraphenylethene
TTE	Tetrathienylethene
UV	Ultraviolet
V	Volt
VB	Valence band
vis	Visible
W	Watt

7 BIBLIOGRAPHY

[1] https://www.bp.com/content/dam/bp/business-sites/en/global/corporate/pdfs/energy-economics/statistical-review/bp-stats-review-2022-full-report.pdf, 71 ed., BP, **2022** (Acces date 01.11.2022).

[2] https://www.bp.com/content/dam/bp/business-sites/en/global/corporate/pdfs/energy-economics/statistical-review/bp-stats-review-2020-full-report.pdf, 69 ed., BP, **2020** (Access date 01.11.2022).

[3] https://www.webofscience.com/wos/woscc/analyze-results/5a2125d0-7853-4836-82d9-8f28351dbaf5-5a8f5457, Web of Science, **2022** (Access date 01.11.2022).

[4] A. E. Becquerel, *Mémoire sur les effets électriques produits sous l'influence des rayons solaires, Vol. 9*, Comptes Rendues, **1839**.

[5] P. Würfel, U. Würfel, *Physics of solar cells : from basic principles to advanced concepts*, **2009**.

[6] W. van Roosbroeck, W. Shockley, *Phys. Rev.* **1954**, *94*, 1558-1560.

[7] W. P. Dumke, *Phys. Rev.* **1957**, *105*, 139-144.

[8] W. Shockley, W. T. Read, *Phys. Rev.* **1952**, *87*.

[9] R. N. Hall, *Phys. Rev.* **1952**, *87*.

[10] P. Auger, *C.R.A.S.* **1923**, *177*.

[11] Y. Abou Jieb, E. Hossain, *Photovoltaic Systems : Fundamentals and Applications*, **2022**.

[12] W. G. Adams, R. E. Day, *Proc. R. Soc. Lond.* **1877**, *25*, 113-117.

[13] W. Smith, *Nature* **1873**, *7*, 303-303.

[14] W. Hallwachs, *Phys. Z.* **1904**, *5*, 489-499.

[15] H. Hertz, *Ann. Phys.* **1887**, *267*, 983-1000.

[16] A. Einstein, *Phys. Z.* **1917**, *18*, 121-136.

[17] F. Bloch, *Z. Phys.* **1929**, *52*, 555-600.

[18] D. M. Chapin, C. S. Fuller, G. L. Pearson, *J. Appl. Phys.* **1954**, *25*, 676-677.

[19] L. M. Fraas, *7Low-Cost Solar Electric Power*, **2014**.

[20] A. W. Blakers, A. Wang, A. M. Milne, J. Zhao, M. A. Green, *Appl. Phys. Lett.* **1989**, *55*, 1363-1365.

[21] J. Zhao, A. Wang, M. A. Green, F. Ferrazza, *Appl. Phys. Lett.* **1998**, *73*, 1991-1993.

[22] W. Shockley, H. J. Queisser, *J. Appl. Phys.* **1961**, *32*, 510-519.

[23] S. Rühle, *Sol. Energy* **2016**, *130*, 139-147.

[24] Z. I. Alferov, V. M. Andreev, M. B. Kagan, I. I. Protasov, V. G. Trofim, *Sov. Phys Semicond.* **1971**, *4*, 2047.

[25] H. J. Hovel, J. M. Woodall, *Appl. Phys. Lett.* **1972**, *21*, 379-381.

[26] B. M. Kayes, H. Nie, R. Twist, S. G. Spruytte, F. Reinhardt, I. C. Kizilyalli, G. S. Higashi, IEEE, **2011**, pp. 04-08.

[27] M. Nakamura, K. Yamaguchi, Y. Kimoto, Y. Yasaki, T. Kato, H. Sugimoto, *IEEE J. Photovolt.* **2019**, *9*, 1863-1867.

[28] J. F. Geisz, R. M. France, K. L. Schulte, M. A. Steiner, A. G. Norman, H. L. Guthrey, M. R. Young, T. Song, T. Moriarty, *Nat. Energy* **2020**, *5*, 326-335.

[29] D. E. Carlson, C. R. Wronski, *Appl. Phys. Lett.* **1976**, *28*, 671-673.

[30]   H. Mette, H. Pick, *Z. Phys.* **1953**, *134*, 566-575.

[31]   H. Akamatu, H. Inokuchi, Y. Matsunaga, *Bull. Chem. Soc. Jpn.* **1956**, *29*, 213-218.

[32]   D. Kearns, M. Calvin, *J. Chem. Phys.* **1958**, *29*, 950-951.

[33]   H. Kallmann, M. Pope, *J. Chem. Phys.* **1959**, *30*, 585-586.

[34]   C. K. Chiang, C. R. Fincher, Y. W. Park, A. J. Heeger, H. Shirakawa, E. J. Louis, S. C. Gau, A. G. MacDiarmid, *Phys. Rev. Lett.* **1977**, *39*, 1098-1101.

[35]   C. W. Tang, *Appl. Phys. Lett.* **1986**, *48*, 183-185.

[36]   Y. Cui, Y. Xu, H. Yao, P. Bi, L. Hong, J. Zhang, Y. Zu, T. Zhang, J. Qin, J. Ren, Z. Chen, C. He, X. Hao, Z. Wei, J. Hou, *Adv. Mater.* **2021**, *33*, 2102420.

[37]   M. B. Upama, M. A. Mahmud, G. Conibeer, A. Uddin, *Sol. RRL* **2020**, *4*, 1900342.

[38]   L. Zhan, S. Li, T.-K. Lau, Y. Cui, X. Lu, M. Shi, C.-Z. Li, H. Li, J. Hou, H. Chen, *Energ. Environ. Sci.* **2020**, *13*, 635-645.

[39]   U. Würfel, J. Herterich, M. List, J. Faisst, M. F. M. Bhuyian, H.-F. Schleiermacher, K. T. Knupfer, B. Zimmermann, *Sol. RRL* **2021**, *5*, 2000802.

[40]   W. Tress, *Organic Solar Cells : Theory, Experiment, and Device Simulation*, **2014**.

[41]   A. Colsmann, H. Röhm, C. Sprau, *Sol. RRL* **2020**, *4*, 2000015.

[42]   B. O'Regan, M. Grätzel, *Nature* **1991**, *353*, 737-740.

[43]   P. Wang, L. Yang, H. Wu, Y. Cao, J. Zhang, N. Xu, S. Chen, J.-D. Decoppet, S. M. Zakeeruddin, M. Grätzel, *Joule* **2018**, *2*, 2145-2153.

[44]   H. Wu, X. Xie, Y. Mei, Y. Ren, Z. Shen, S. Li, P. Wang, *ACS Photonics* **2019**, *6*, 1216-1225.

[45]   S. Mathew, A. Yella, P. Gao, R. Humphry-Baker, B. F. E. Curchod, N. Ashari-Astani, I. Tavernelli, U. Rothlisberger, M. K. Nazeeruddin, M. Grätzel, *Nat. Chem.* **2014**, *6*, 242-247.

[46]   B. Conings, L. Baeten, T. Jacobs, R. Dera, J. D'Haen, J. Manca, H. G. Boyen, *APL Mater.* **2014**, *2*, 081505.

[47]   S. Ahmad, S. Kazim, M. Grätzel, *Perovskite solar cells : materials, processes, and devices*, **2022**.

[48]   J. Urieta-Mora, I. García-Benito, A. Molina-Ontoria, N. Martín, *Chem. Soc. Rev.* **2018**, *47*, 8541-8571.

[49]   A. Kojima, K. Teshima, Y. Shirai, T. Miyasaka, *J. Am. Chem. Soc.* **2009**, *131*, 6050-6051.

[50]   J.-H. Im, C.-R. Lee, J.-W. Lee, S.-W. Park, N.-G. Park, *Nanoscale* **2011**, *3*, 4088-4093.

[51]   H.-S. Kim, C.-R. Lee, J.-H. Im, K.-B. Lee, T. Moehl, A. Marchioro, S.-J. Moon, R. Humphry-Baker, J.-H. Yum, J. E. Moser, M. Grätzel, N.-G. Park, *Sci. Rep.* **2012**, *2*, 591.

[52]   M. M. Lee, J. Teuscher, T. Miyasaka, T. N. Murakami, H. J. Snaith, *Science* **2012**, *338*, 643-647.

[53]   Q. Jiang, Y. Zhao, X. Zhang, X. Yang, Y. Chen, Z. Chu, Q. Ye, X. Li, Z. Yin, J. You, *Nat. Photon.* **2019**, *13*, 460-466.

[54]   Y. Gao, E. Shi, S. Deng, S. B. Shiring, J. M. Snaider, C. Liang, B. Yuan, R. Song, S. M. Janke, A. Liebman-Peláez, P. Yoo, M. Zeller, B. W. Boudouris, P. Liao, C. Zhu, V. Blum, Y. Yu, B. M. Savoie, L. Huang, L. Dou, *Nat. Chem.* **2019**, *11*, 1151-1157.

[55]   M. Ikram, R. Malik, R. Raees, M. Imran, F. Wang, S. Ali, M. Khan, Q. Khan, M. Maqbool, *Sustainable Energy Technologies and Assessments* **2022**, *53*, 102433.

[56]   https://www.helmholtz-berlin.de/pubbin/news_seite?nid=21020;sprache=en;seitenid=1, **2020** (Access date 10.06.2020).

[57]   https://actu.epfl.ch/news/new-world-records-perovskite-on-silicon-tandem-sol/, **2022** (Access date 03.11.2022).

226

[58] S. Gharibzadeh, I. M. Hossain, P. Fassl, B. A. Nejand, T. Abzieher, M. Schultes, E. Ahlswede, P. Jackson, M. Powalla, S. Schäfer, M. Rienäcker, T. Wietler, R. Peibst, U. Lemmer, B. S. Richards, U. W. Paetzold, *Adv. Funct. Mater.* **2020**, *30*, 1909919.

[59] Y. Nishigaki, T. Nagai, M. Nishiwaki, T. Aizawa, M. Kozawa, K. Hanzawa, Y. Kato, H. Sai, H. Hiramatsu, H. Hosono, H. Fujiwara, *Sol. RRL* **2020**, *4*, 1900555.

[60] https://www.nrel.gov/pv/cell-efficiency.html, **2022** (Access date 01.11.2022).

[61] J. Lian, B. Lu, F. Niu, P. Zeng, X. Zhan, *Small Methods* **2018**, *2*, 1800082.

[62] H. S. Jung, N.-G. Park, *Small* **2015**, *11*, 10-25.

[63] F. Arntz, Y. Yacoby, *Phys. Rev. Lett.* **1966**, *17*, 857-860.

[64] H. Liu, Z. Huang, S. Wei, L. Zheng, L. Xiao, Q. Gong, *Nanoscale* **2016**, *8*, 6209-6221.

[65] H. Lu, J. Zhong, C. Ji, J. Zhao, D. Li, R. Zhao, Y. Jiang, S. Fang, T. Liang, H. Li, C. M. Li, *Nano Energy* **2020**, *68*, 104336.

[66] S. Foo, M. Thambidurai, P. Senthil Kumar, R. Yuvakkumar, Y. Huang, C. Dang, *Int. J. Energy Res.* **2022**, *46*, 21441 - 21451.

[67] Y. Tang, R. Roy, Z. Zhang, Y. Hu, F. Yang, C. Qin, L. Jiang, H. Liu, *Sol. Energy* **2022**, *231*, 440-446.

[68] Y. Sanehira, N. Shibayama, Y. Numata, M. Ikegami, T. Miyasaka, *ACS Appl. Mater. Interfaces* **2020**, *12*, 15175-15182.

[69] H. Tan, A. Jain, O. Voznyy, X. Lan, F. P. García de Arquer, J. Z. Fan, R. Quintero-Bermudez, M. Yuan, B. Zhang, Y. Zhao, F. Fan, P. Li, L. N. Quan, Y. Zhao, Z.-H. Lu, Z. Yang, S. Hoogland, E. H. Sargent, *Science* **2017**, *355*, 722-726.

[70] L. Lin, T. W. Jones, T. C.-J. Yang, N. W. Duffy, J. Li, L. Zhao, B. Chi, X. Wang, G. J. Wilson, *Adv. Funct. Mater.* **2021**, *31*, 2008300.

[71] L. Xiong, Y. Guo, J. Wen, H. Liu, G. Yang, P. Qin, G. Fang, *Adv. Funct. Mater.* **2018**, *28*, 1802757.

[72] Q. Jiang, X. Zhang, J. You, *Small* **2018**, *14*, 1801154.

[73] A. S. Subbiah, N. Mathews, S. Mhaisalkar, S. K. Sarkar, *ACS Energy Lett.* **2018**, *3*, 1482-1491.

[74] Q. Guo, J. Wu, Y. Yang, X. Liu, Z. Lan, J. Lin, M. Huang, Y. Wei, J. Dong, J. Jia, Y. Huang, *Research* **2019**, *2019*, 4049793.

[75] D. Yang, R. Yang, K. Wang, C. Wu, X. Zhu, J. Feng, X. Ren, G. Fang, S. Priya, S. Liu, *Nat. Commun.* **2018**, *9*, 3239.

[76] D. Zheng, G. Wang, W. Huang, B. Wang, W. Ke, J. L. Logsdon, H. Wang, Z. Wang, W. Zhu, J. Yu, M. R. Wasielewski, M. G. Kanatzidis, T. J. Marks, A. Facchetti, *Adv. Funct. Mater.* **2019**, *29*, 1900265.

[77] L. Zhu, Z. Shao, J. Ye, X. Zhang, X. Pan, S. Dai, *Chem. Commun.* **2016**, *52*, 970-973.

[78] J. Liu, C. Gao, L. Luo, Q. Ye, X. He, L. Ouyang, X. Guo, D. Zhuang, C. Liao, J. Mei, W. Lau, *J. Mater. Chem. A* **2015**, *3*, 11750-11755.

[79] F. Jafari, M. Ahmadpour, U. K. Aryal, M. Ahmad, M. Prete, N. Torabi, V. Turkovic, H.-G. Rubahn, A. Behjat, M. Madsen, in *Perovskite Solar Cells*, **2021**, pp. 311-329.

[80] S.-M. Dai, H.-R. Tian, M.-L. Zhang, Z. Xing, L.-Y. Wang, X. Wang, T. Wang, L.-L. Deng, S.-Y. Xie, R.-B. Huang, L.-S. Zheng, *J. Power Sources* **2017**, *339*, 27-32.

[81] J.-Y. Jeng, Y.-F. Chiang, M.-H. Lee, S.-R. Peng, T.-F. Guo, P. Chen, T.-C. Wen, *Adv. Mater.* **2013**, *25*, 3727-3732.

[82]   D. Liu, Q. Wang, C. J. Traverse, C. Yang, M. Young, P. S. Kuttipillai, S. Y. Lunt, T. W. Hamann, R. R. Lunt, *ACS Nano* **2018**, *12*, 876-883.

[83]   P. Docampo, J. M. Ball, M. Darwich, G. E. Eperon, H. J. Snaith, *Nat. Commun.* **2013**, *4*, 2761.

[84]   X. Yin, J. Zhai, L. Song, P. Du, N. Li, Y. Yang, J. Xiong, F. Ko, *ACS Appl. Mater. Interfaces* **2019**, *11*, 44308-44314.

[85]   I. M. Chan, T.-Y. Hsu, F. C. Hong, *Appl. Phys. Lett.* **2002**, *81*, 1899-1901.

[86]   K.-C. Wang, J.-Y. Jeng, P.-S. Shen, Y.-C. Chang, E. W.-G. Diau, C.-H. Tsai, T.-Y. Chao, H.-C. Hsu, P.-Y. Lin, P. Chen, T.-F. Guo, T.-C. Wen, *Sci. Rep.* **2014**, *4*, 4756.

[87]   Q. Wang, C.-C. Chueh, T. Zhao, J. Cheng, M. Eslamian, W. C. H. Choy, A. K. Y. Jen, *ChemSusChem* **2017**, *10*, 3794-3803.

[88]   L. J. Tang, X. Chen, T. Y. Wen, S. Yang, J. J. Zhao, H. W. Qiao, Y. Hou, H. G. Yang, *Chem. Eur. J.* **2018**, *24*, 2845-2849.

[89]   K. Yao, F. Li, Q. He, X. Wang, Y. Jiang, H. Huang, A. K. Y. Jen, *Nano Energy* **2017**, *40*, 155-162.

[90]   J. Ma, M. Zheng, C. Chen, Z. Zhu, X. Zheng, Z. Chen, Y. Guo, C. Liu, Y. Yan, G. Fang, *Adv. Funct. Mater.* **2018**, *28*, 1804128.

[91]   Y. Xie, K. Lu, J. Duan, Y. Jiang, L. Hu, T. Liu, Y. Zhou, B. Hu, *ACS Appl. Mater. Interfaces* **2018**, *10*, 14153-14159.

[92]   W. Chen, F.-Z. Liu, X.-Y. Feng, A. B. Djurišić, W. K. Chan, Z.-B. He, *Adv. Energy Mater.* **2017**, *7*, 1700722.

[93]   X. Wan, Y. Jiang, Z. Qiu, H. Zhang, X. Zhu, I. Sikandar, X. Liu, X. Chen, B. Cao, *ACS Appl. Energy Mater.* **2018**, *1*, 3947-3954.

[94]   H. Zhang, H. Wang, H. Zhu, C.-C. Chueh, W. Chen, S. Yang, A. K. Y. Jen, *Adv. Energy Mater.* **2018**, *8*, 1702762.

[95]   C. Zuo, L. Ding, *Small* **2015**, *11*, 5528-5532.

[96]   H. Rao, S. Ye, W. Sun, W. Yan, Y. Li, H. Peng, Z. Liu, Z. Bian, Y. Li, C. Huang, *Nano Energy* **2016**, *27*, 51-57.

[97]   H. Rao, W. Sun, S. Ye, W. Yan, Y. Li, H. Peng, Z. Liu, Z. Bian, C. Huang, *ACS Appl. Mater. Interfaces* **2016**, *8*, 7800-7805.

[98]   H. Wang, Z. Yu, J. Lai, X. Song, X. Yang, A. Hagfeldt, L. Sun, *J. Mater. Chem. A* **2018**, *6*, 21435-21444.

[99]   Z.-L. Tseng, L.-C. Chen, C.-H. Chiang, S.-H. Chang, C.-C. Chen, C.-G. Wu, *Sol. Energy* **2016**, *139*, 484-488.

[100]   P. Schulz, J. O. Tiepelt, J. A. Christians, I. Levine, E. Edri, E. M. Sanehira, G. Hodes, D. Cahen, A. Kahn, *ACS Appl. Mater. Interfaces* **2016**, *8*, 31491-31499.

[101]   F. Hou, Z. Su, F. Jin, X. Yan, L. Wang, H. Zhao, J. Zhu, B. Chu, W. Li, *Nanoscale* **2015**, *7*, 9427-9432.

[102]   D. B. Khadka, Y. Shirai, M. Yanagida, J. W. Ryan, K. Miyano, *J. Mater. Chem. C* **2017**, *5*, 8819-8827.

[103]   D. Liu, Y. Li, J. Yuan, Q. Hong, G. Shi, D. Yuan, J. Wei, C. Huang, J. Tang, M.-K. Fung, *J. Mater. Chem. A* **2017**, *5*, 5701-5708.

[104]   C. Zuo, L. Ding, *Adv. Energy Mater.* **2017**, *7*, 1601193.

[105]   X. Zhou, J. Blochwitz, M. Pfeiffer, A. Nollau, T. Fritz, K. Leo, *Adv. Funct. Mater.* **2001**, *11*, 310-314.

[106]   K.-G. Lim, S. Ahn, Y.-H. Kim, Y. Qi, T.-W. Lee, *Energ. Environ. Sci.* **2016**, *9*, 932-939.

[107] P. Tehrani, A. Kanciurzewska, X. Crispin, N. D. Robinson, M. Fahlman, M. Berggren, *Solid State Ion.* **2007**, *177*, 3521-3527.

[108] Q. Wang, C.-C. Chueh, M. Eslamian, A. K. Y. Jen, *ACS Appl. Mater. Interfaces* **2016**, *8*, 32068-32076.

[109] Y. Wang, Y. Hu, D. Han, Q. Yuan, T. Cao, N. Chen, D. Zhou, H. Cong, L. Feng, *Org. Electron.* **2019**, *70*, 63-70.

[110] M. Cui, H. Rui, X. Wu, Z. Sun, W. Qu, W. Qin, S. Yin, *J. Phys. Chem. Lett.* **2021**, *12*, 8533-8540.

[111] W. Zhang, J. Smith, R. Hamilton, M. Heeney, J. Kirkpatrick, K. Song, S. E. Watkins, T. Anthopoulos, I. McCulloch, *J. Am. Chem. Soc.* **2009**, *131*, 10814-10815.

[112] Q. Wang, Q. Dong, T. Li, A. Gruverman, J. Huang, *Adv. Mater.* **2016**, *28*, 6734-6739.

[113] J. Xu, J. Dai, H. Dong, P. Li, J. Chen, X. Zhu, Z. Wang, B. Jiao, X. Hou, J. Li, Z. Wu, *Org. Electron.* **2022**, *100*, 106378.

[114] B. Fan, Z. He, J. Xiong, Q. Zhao, Z. Dai, Z. Zhang, X. Xue, J. Yang, B. Yang, J. Zhang, *Org. Electron.* **2019**, *75*, 105396.

[115] D. Luo, W. Yang, Z. Wang, A. Sadhanala, Q. Hu, R. Su, R. Shivanna, G. F. Trindade, J. F. Watts, Z. Xu, T. Liu, K. Chen, F. Ye, P. Wu, L. Zhao, J. Wu, Y. Tu, Y. Zhang, X. Yang, W. Zhang, R. H. Friend, Q. Gong, H. J. Snaith, R. Zhu, *Science* **2018**, *360*, 1442-1446.

[116] S. Tian, J. Chen, X. Lian, Y. Wang, Y. Zhang, W. Yang, G. Wu, W. Qiu, H. Chen, *J. Mater. Chem. A* **2019**, *7*, 14027-14032.

[117] J. Wang, J. Xu, Z. Li, X. Lin, C. Yu, H. Wu, H.-l. Wang, *ACS Appl. Energy Mater.* **2020**, *3*, 6344-6351.

[118] D. Zhao, M. Sexton, H.-Y. Park, G. Baure, J. C. Nino, F. So, *Adv. Energy Mater.* **2015**, *5*, 1401855.

[119] X. Xu, C. Ma, Y. Cheng, Y.-M. Xie, X. Yi, B. Gautam, S. Chen, H.-W. Li, C.-S. Lee, F. So, S.-W. Tsang, *J. Power Sources* **2017**, *360*, 157-165.

[120] F. Zhang, Y. Hou, S. Wang, H. Zhang, F. Zhou, Y. Hao, S. Ye, H. Cai, J. Song, J. Qu, *Sol. RRL* **2021**, *5*, 2100190.

[121] B. Conings, L. Baeten, C. De Dobbelaere, J. D'Haen, J. Manca, H.-G. Boyen, *Adv. Mater.* **2014**, *26*, 2041-2046.

[122] Y. Zhang, M. Elawad, Z. Yu, X. Jiang, J. Lai, L. Sun, *RSC Adv.* **2016**, *6*, 108888-108895.

[123] P. Zhou, T. Bu, S. Shi, L. Li, Y. Zhang, Z. Ku, Y. Peng, J. Zhong, Y.-B. Cheng, F. Huang, *J. Mater. Chem. C* **2018**, *6*, 5733-5737.

[124] E. H. Jung, N. J. Jeon, E. Y. Park, C. S. Moon, T. J. Shin, T.-Y. Yang, J. H. Noh, J. Seo, *Nature* **2019**, *567*, 511-515.

[125] X. Li, X. Liu, X. Wang, L. Zhao, T. Jiu, J. Fang, *J. Mater. Chem. A* **2015**, *3*, 15024-15029.

[126] X. Li, Y.-C. Wang, L. Zhu, W. Zhang, H.-Q. Wang, J. Fang, *ACS Appl. Mater. Interfaces* **2017**, *9*, 31357-31361.

[127] S. Li, B. He, J. Xu, H. Lu, J. Jiang, J. Zhu, Z. Kan, L. Zhu, F. Wu, *Nanoscale* **2020**, *12*, 3686-3691.

[128] Q. Liao, Y. Wang, X. Yao, M. Su, B. Li, H. Sun, J. Huang, X. Guo, *ACS Appl. Mater. Interfaces* **2021**, *13*, 16744-16753.

[129] J. Salbeck, N. Yu, J. Bauer, F. Weissörtel, H. Bestgen, *Synth. Met.* **1997**, *91*, 209-215.

[130] U. Bach, D. Lupo, P. Comte, J. E. Moser, F. Weissörtel, J. Salbeck, H. Spreitzer, M. Grätzel, *Nature* **1998**, *395*, 583-585.

[131] Z. Hu, W. Fu, L. Yan, J. Miao, H. Yu, Y. He, O. Goto, H. Meng, H. Chen, W. Huang, *Chem. Sci.* **2016**, *7*, 5007-5012.

[132] N. J. Jeon, H. G. Lee, Y. C. Kim, J. Seo, J. H. Noh, J. Lee, S. I. Seok, *J. Am. Chem. Soc.* **2014**, *136*, 7837-7840.

[133] A. T. Murray, J. M. Frost, C. H. Hendon, C. D. Molloy, D. R. Carbery, A. Walsh, *Chem. Commun.* **2015**, *51*, 8935-8938.

[134] H. J. Snaith, M. Grätzel, *Appl. Phys. Lett.* **2006**, *89*, 262114.

[135] A. Abate, T. Leijtens, S. Pathak, J. Teuscher, R. Avolio, M. E. Errico, J. Kirkpatrik, J. M. Ball, P. Docampo, I. McPherson, H. J. Snaith, *Phys. Chem. Chem. Phys.* **2013**, *15*, 2572-2579.

[136] I. Lee, J. H. Yun, H. J. Son, T.-S. Kim, *ACS Appl. Mater. Interfaces* **2017**, *9*, 7029-7035.

[137] A. K. Jena, M. Ikegami, T. Miyasaka, *ACS Energy Lett.* **2017**, *2*, 1760-1761.

[138] S. Wang, Z. Huang, X. Wang, Y. Li, M. Günther, S. Valenzuela, P. Parikh, A. Cabreros, W. Xiong, Y. S. Meng, *J. Am. Chem. Soc.* **2018**, *140*, 16720-16730.

[139] E. J. Juarez-Perez, M. R. Leyden, S. Wang, L. K. Ono, Z. Hawash, Y. Qi, *Chem. Mater.* **2016**, *28*, 5702-5709.

[140] S. Wang, M. Sina, P. Parikh, T. Uekert, B. Shahbazian, A. Devaraj, Y. S. Meng, *Nano Lett.* **2016**, *16*, 5594-5600.

[141] S. N. Habisreutinger, N. K. Noel, H. J. Snaith, R. J. Nicholas, *Adv. Energy Mater.* **2017**, *7*, 1601079.

[142] B. Tan, S. R. Raga, A. S. R. Chesman, S. O. Fürer, F. Zheng, D. P. McMeekin, L. Jiang, W. Mao, X. Lin, X. Wen, J. Lu, Y.-B. Cheng, U. Bach, *Adv. Energy Mater.* **2019**, *9*, 1901519.

[143] G. Ren, W. Han, Y. Deng, W. Wu, Z. Li, J. Guo, H. Bao, C. Liu, W. Guo, *J. Mater. Chem. A* **2021**, *9*, 4589-4625.

[144] A. D. Scaccabarozzi, A. Basu, F. Aniés, J. Liu, O. Zapata-Arteaga, R. Warren, Y. Firdaus, M. I. Nugraha, Y. Lin, M. Campoy-Quiles, N. Koch, C. Müller, L. Tsetseris, M. Heeney, T. D. Anthopoulos, *Chem. Rev.* **2022**, *122*, 4420-4492.

[145] C. Poriel, Y. Ferrand, S. Juillard, P. Le Maux, G. Simonneaux, *Tetrahedron* **2004**, *60*, 145-158.

[146] M. Saliba, S. Orlandi, T. Matsui, S. Aghazada, M. Cavazzini, J.-P. Correa-Baena, P. Gao, R. Scopelliti, E. Mosconi, K.-H. Dahmen, F. De Angelis, A. Abate, A. Hagfeldt, G. Pozzi, M. Grätzel, M. K. Nazeeruddin, *Nat. Energy* **2016**, *1*, 15017.

[147] I. M. Abdellah, T. H. Chowdhury, J.-J. Lee, A. Islam, M. K. Nazeeruddin, M. Grätzel, A. El-Shafei, *Sustain. Energy Fuels* **2021**, *5*, 199-211.

[148] S. Kazim, F. J. Ramos, P. Gao, M. K. Nazeeruddin, M. Grätzel, S. Ahmad, *Energ. Environ. Sci.* **2015**, *8*, 1816-1823.

[149] H. D. Pham, H. Hu, F.-L. Wong, C.-S. Lee, W.-C. Chen, K. Feron, S. Manzhos, H. Wang, N. Motta, Y. M. Lam, P. Sonar, *J. Mater. Chem. C* **2018**, *6*, 9017-9029.

[150] Q.-Q. Ge, J.-Y. Shao, J. Ding, L.-Y. Deng, W.-K. Zhou, Y.-X. Chen, J.-Y. Ma, L.-J. Wan, J. Yao, J.-S. Hu, Y.-W. Zhong, *Angew. Chem. Int. Ed.* **2018**, *57*, 10959-10965.

[151] J. Qiu, H. Liu, X. Li, S. Wang, *Chem. Eng. J.* **2020**, *387*, 123965.

[152] Y.-C. Chang, K.-M. Lee, C.-H. Lai, C.-Y. Liu, *Chem. Asian J.* **2018**, *13*, 1510-1515.

[153] H. Choi, H. M. Ko, J. Ko, *Dyes Pigm.* **2016**, *126*, 179-185.

[154] H. Choi, S. Park, S. Paek, P. Ekanayake, M. K. Nazeeruddin, J. Ko, *J. Mater. Chem. A* **2014**, *2*, 19136-19140.

[155] C. Liu, D. Zhang, Z. Li, X. Zhang, L. Shen, W. Guo, *Adv. Funct. Mater.* **2018**, *28*, 1803126.

[156]   A.-N. Cho, H.-S. Kim, T.-T. Bui, X. Sallenave, F. Goubard, N.-G. Park, *RSC Adv.* **2016**, *6*, 68553-68559.

[157]   C. J. Brown, A. C. Farthing, *Nature* **1949**, *164*, 915-916.

[158]   D. J. Cram, H. Steinberg, *J. Am. Chem. Soc.* **1951**, *73*, 5691-5704.

[159]   Z. Hassan, E. Spuling, D. M. Knoll, J. Lahann, S. Bräse, *Chem. Soc. Rev.* **2018**, *47*, 6947-6963.

[160]   Z. Hassan, E. Spuling, D. M. Knoll, S. Brase, *Angew. Chem. Int. Ed.* **2020**, *59*, 2156-2170.

[161]   E. Spuling, N. Sharma, I. D. W. Samuel, E. Zysman-Colman, S. Bräse, *Chem. Commun.* **2018**, *54*, 9278-9281.

[162]   N. Sharma, E. Spuling, C. M. Mattern, W. Li, O. Fuhr, Y. Tsuchiya, C. Adachi, S. Bräse, I. D. W. Samuel, E. Zysman-Colman, *Chem. Sci.* **2019**, *10*, 6689-6696.

[163]   S. Park, J. H. Heo, C. H. Cheon, H. Kim, S. H. Im, H. J. Son, *J. Mater. Chem. A* **2015**, *3*, 24215-24220.

[164]   S. Park, J. H. Heo, J. H. Yun, T. S. Jung, K. Kwak, M. J. Ko, C. H. Cheon, J. Y. Kim, S. H. Im, H. J. Son, *Chem. Sci.* **2016**, *7*, 5517-5522.

[165]   Y.-S. Lin, H. Li, W.-S. Yu, S.-T. Wang, Y.-M. Chang, T.-H. Liu, S.-S. Li, M. Watanabe, H.-H. Chiu, D.-Y. Wang, Y. J. Chang, *J. Power Sources* **2021**, *491*, 229543.

[166]   W. Ke, P. Priyanka, S. Vegiraju, C. C. Stoumpos, I. Spanopoulos, C. M. M. Soe, T. J. Marks, M.-C. Chen, M. G. Kanatzidis, *J. Am. Chem. Soc.* **2018**, *140*, 388-393.

[167]   C. Shen, Y. Wu, H. Zhang, E. Li, W. Zhang, X. Xu, W. Wu, H. Tian, W.-H. Zhu, *Angew. Chem. Int. Ed.* **2019**, *58*, 3784-3789.

[168]   X. Zhang, Z. Zhou, S. Ma, G. Wu, X. Liu, M. Mateen, R. Ghadari, Y. Wu, Y. Ding, M. Cai, S. Dai, *Chem. Commun.* **2020**, *56*, 3159-3162.

[169]   X. Zhang, X. Liu, R. Ghadari, M. Li, Z. Zhou, Y. Ding, M. Cai, S. Dai, *ACS Appl. Mater. Interfaces* **2021**, *13*, 12322-12330.

[170]   P. Rothemund, A. R. Menotti, *J. Am. Chem. Soc.* **1941**, *63*, 267-270.

[171]   A. D. Adler, F. R. Longo, W. Shergalis, *J. Am. Chem. Soc.* **1964**, *86*, 3145-3149.

[172]   T. J. Savenije, A. Goossens, *Phys. Rev. B* **2001**, *64*, 115323.

[173]   F. Hernández-Fernández, M. Pavanello, L. Visscher, *Phys. Chem. Chem. Phys.* **2016**, *18*, 21122-21132.

[174]   S. Chen, P. Liu, Y. Hua, Y. Li, L. Kloo, X. Wang, B. Ong, W.-K. Wong, X. Zhu, *ACS Appl. Mater. Interfaces* **2017**, *9*, 13231-13239.

[175]   U.-H. Lee, R. Azmi, S. Sinaga, S. Hwang, S. H. Eom, T.-W. Kim, S. C. Yoon, S.-Y. Jang, I. H. Jung, *ChemSusChem* **2017**, *10*, 3780-3787.

[176]   R. Azmi, U.-H. Lee, F. T. A. Wibowo, S. H. Eom, S. C. Yoon, S.-Y. Jang, I. H. Jung, *ACS Appl. Mater. Interfaces* **2018**, *10*, 35404-35410.

[177]   S. H. Kang, C. Lu, H. Zhou, S. Choi, J. Kim, H. K. Kim, *Dyes Pigm.* **2019**, *163*, 734-739.

[178]   H.-H. Chou, Y.-H. Chiang, Y.-H. Chen, C.-J. Guo, H.-Y. Zuo, W.-T. Cheng, P.-Y. Lin, Y. Y. Chiu, P. Chen, C.-Y. Yeh, *Sol. RRL* **2020**, *4*, 2000119.

[179]   T. Sun, Z. Zhang, J. Xu, L. Liang, C.-L. Mai, L. Ren, Q. Zhou, Y. Yu, B. Zhang, P. Gao, *Dyes Pigm.* **2021**, *193*, 109469.

[180]   F. Ullmann, *Ber. Dtsch. Chem. Ges.* **1903**, *36*, 2382-2384.

[181]   I. Goldberg, *Ber. Dtsch. Chem. Ges.* **1906**, *39*, 1691-1692.

[182]   S. V. Ley, A. W. Thomas, *Angew. Chem. Int. Ed.* **2003**, *42*, 5400-5449.

[183]   H. B. Goodbrand, N.-X. Hu, *J. Org. Chem.* **1999**, *64*, 670-674.

231

[184]   T.-Y. Li, C. Su, S. B. Akula, W.-G. Sun, H.-M. Chien, W.-R. Li, *Org. Lett.* **2016**, *18*, 3386-3389.

[185]   D. Wright, U. Gubler, W. E. Moerner, M. S. DeClue, J. S. Siegel, *J. Phys. Chem. B* **2003**, *107*, 4732-4737.

[186]   J. Safaei-Ghomi, Z. Akbarzadeh, A. ziarati, *RSC Adv.* **2014**, *4*, 16385-16390.

[187]   S. H. Kim, J. Choi, C. Sakong, J. W. Namgoong, W. Lee, D. H. Kim, B. Kim, M. J. Ko, J. P. Kim, *Dyes Pigm.* **2015**, *113*, 390-401.

[188]   I. Lee, J. E. Kwon, Y. Kang, K. C. Kim, B.-G. Kim, *ACS Sensors* **2018**, *3*, 1831-1837.

[189]   D. Lumpi, B. Holzer, J. Bintinger, E. Horkel, S. Waid, H. D. Wanzenböck, M. Marchetti-Deschmann, C. Hametner, E. Bertagnolli, I. Kymissis, J. Fröhlich, *New J. Chem.* **2015**, *39*, 1840-1851.

[190]   J. Niu, P. Guo, J. Kang, Z. Li, J. Xu, S. Hu, *J. Org. Chem.* **2009**, *74*, 5075-5078.

[191]   M. Holzapfel, C. Lambert, *J. Phys. Chem. C* **2008**, *112*, 1227-1243.

[192]   M. Mao, M.-G. Ren, Q.-H. Song, *Chem. Eur. J.* **2012**, *18*, 15512-15522.

[193]   L. Wang, E. Ji, N. Liu, B. Dai, *Synthesis* **2016**, *48*, 737-750.

[194]   F. Vögtle, P. Neumann, *Tetrahedron Lett.* **1969**, *10*, 5329-5334.

[195]   F. Salhi, B. Lee, C. Metz, L. A. Bottomley, D. M. Collard, *Org. Lett.* **2002**, *4*, 3195-3198.

[196]   F. Salhi, D. M. Collard, *Adv. Mater.* **2003**, *15*, 81-85.

[197]   I. Dix, H. Hopf, P. G. Jones, *Acta Cryst.* **1999**, *C55*, 756-759.

[198]   D. Milstein, J. K. Stille, *J. Am. Chem. Soc.* **1978**, *100*, 3636-3638.

[199]   K. J. Weiland, N. Münch, W. Gschwind, D. Häussinger, M. Mayor, *Helv. Chim. Acta* **2019**, *102*, e1800205.

[200]   K. Tamao, K. Sumitani, M. Kumada, *J. Am. Chem. Soc.* **1972**, *94*, 4374-4376.

[201]   V. I. Rozenberg, E. V. Sergeeva, V. G. Kharitonov, N. V. Vorontsova, E. V. Vorontsov, V. V. Mikul'shina, *Russ. Chem. Bull.* **1994**, *43*, 1018-1023.

[202]   Y. Akita, Y. Itagaki, S. Takizawa, A. Ohta, *Chem. Pharm. Bull.* **1989**, *37*, 1477-1480.

[203]   J. Wencel-Delord, T. Dröge, F. Liu, F. Glorius, *Chem. Soc. Rev.* **2011**, *40*, 4740-4761.

[204]   C.-Y. Liu, H. Zhao, H.-h. Yu, *Org. Lett.* **2011**, *13*, 4068-4071.

[205]   C. Kieffer, V. Babin, M. Jouanne, I. Slimani, Y. Berhault, R. Legay, J. Sopková-de Oliveira Santos, S. Rault, A. S. Voisin-Chiret, *Tetrahedron* **2017**, *73*, 5509-5516.

[206]   M. Lafrance, K. Fagnou, *J. Am. Chem. Soc.* **2006**, *128*, 16496-16497.

[207]   D. J. Schipper, K. Fagnou, *Chem. Mater.* **2011**, *23*, 1594-1600.

[208]   N. Miyaura, K. Yamada, A. Suzuki, *Tetrahedron Lett.* **1979**, *20*, 3437-3440.

[209]   T. Ishiyama, M. Murata, N. Miyaura, *J. Org. Chem.* **1995**, *60*, 7508-7510.

[210]   S. Barroso, P. Joksch, P. Puylaert, S. Tin, S. J. Bell, L. Donnellan, S. Duguid, C. Muir, P. Zhao, V. Farina, D. N. Tran, J. G. de Vries, *J. Org. Chem.* **2021**, *86*, 103-109.

[211]   K. Matos, J. A. Soderquist, *J. Org. Chem.* **1998**, *63*, 461-470.

[212]   Y. Dienes, S. Durben, T. Kárpáti, T. Neumann, U. Englert, L. Nyulászi, T. Baumgartner, *Chem. Eur. J.* **2007**, *13*, 7487-7500.

[213]   Z. Zhou, X. Zhang, R. Ghadari, X. Liu, W. Wang, Y. Ding, M. Cai, J. H. Pan, S. Dai, *Sol. Energy* **2021**, *221*, 323-331.

[214]   Z. Yao, F. Zhang, H. He, X. Bi, Y. Guo, Y. Guo, L. Wang, X. Wan, Y. Chen, L. Sun, *Angew. Chem. Int. Ed.* **2022**, *61*, e202201847.

[215]   T. Doba, L. Ilies, W. Sato, R. Shang, E. Nakamura, *Nat. Catal.* **2021**, *4*, 631-638.

[216]   M. Fontani, E. Garoni, A. Colombo, C. Dragonetti, S. Fantacci, H. Doucet, J.-F. Soulé, J. Boixel, V. Guerchais, D. Roberto, *Dalton Trans.* **2019**, *48*, 202-208.

[217]   Y. J. Chang, P.-T. Chou, S.-Y. Lin, M. Watanabe, Z.-Q. Liu, J.-L. Lin, K.-Y. Chen, S.-S. Sun, C.-Y. Liu, T. J. Chow, *Chem. Asian J.* **2012**, *7*, 572-581.

[218]   E. M. Barea, C. Zafer, B. Gultekin, B. Aydin, S. Koyuncu, S. Icli, F. F. Santiago, J. Bisquert, *J. Phys. Chem. C* **2010**, *114*, 19840-19848.

[219]   S. K. Jeon, B. Thirupathaiah, C. Kim, K. T. Lim, J. Y. Lee, S. Seo, *Dyes Pigm.* **2015**, *114*, 146-150.

[220]   Q. Feng, X. Jia, G. Zhou, Z.-S. Wang, *Chem. Commun.* **2013**, *49*, 7445-7447.

[221]   S. Amthor, B. Noller, C. Lambert, *Chem. Phys.* **2005**, *316*, 141-152.

[222]   X. Liu, F. Kong, R. Ghadari, S. Jin, W. Chen, T. Yu, T. Hayat, A. Alsaedi, F. Guo, Z. Tan, J. Chen, S. Dai, *Energy Technology* **2017**, *5*, 1788-1794.

[223]   H. Li, K. Fu, A. Hagfeldt, M. Grätzel, S. G. Mhaisalkar, A. C. Grimsdale, *Angew. Chem. Int. Ed.* **2014**, *53*, 4085-4088.

[224]   A. J. Bard, L. R. Faulkner, *Electrochemical Methods: Fundamentals and Applications*, 2 ed., John Wiley & Sons, New York, USA, **2008**.

[225]   X. Liu, F. Kong, R. Ghadari, S. Jin, T. Yu, W. Chen, G. Liu, Z. Tan, J. Chen, S. Dai, *Chem. Commun.* **2017**, *53*, 9558-9561.

[226]   N. Miyaura, A. Suzuki, *Chem. Rev.* **1995**, *95*, 2457-2483.

[227]   J. P. Knowles, A. Whiting, *Org. Biomol. Chem.* **2007**, *5*, 31-44.

[228]   A. O. King, N. Okukado, E.-i. Negishi, *J. Chem. Soc., Chem. Commun.* **1977**, 683-684.

[229]   K. Kobayakawa, M. Hasegawa, H. Sasaki, J. Endo, H. Matsuzawa, K. Sako, J. Yoshida, Y. Mazaki, *Chem. Asian J.* **2014**, *9*, 2751-2754.

[230]   Q. Liu, Y. Lan, J. Liu, G. Li, Y.-D. Wu, A. Lei, *J. Am. Chem. Soc.* **2009**, *131*, 10201-10210.

[231]   L. Jin, A. Lei, *Org. Biomol. Chem.* **2012**, *10*, 6817-6825.

[232]   B. Slimi, M. Mollar, I. B. Assaker, I. Kriaa, R. Chtourou, B. Marí, *Energy Procedia* **2016**, *102*, 87-95.

[233]   M. O. Senge, *Chem. Commun.* **2011**, *47*, 1943-1960.

[234]   C.-W. Huang, K. Yuan Chiu, S.-H. Cheng, *Dalton Trans.* **2005**, 2417-2422.

[235]   S. Fu, X. Zhu, G. Zhou, W.-Y. Wong, C. Ye, W.-K. Wong, Z. Li, *Eur. J. Inorg. Chem.* **2007**, *2007*, 2004-2013.

[236]   Y. Zhang, S. Zhang, L. Yao, L. Qin, L. Nian, Z. Xie, B. Yang, L. Liu, Y. Ma, *Eur. J. Inorg. Chem.* **2014**, *2014*, 4852-4857.

[237]   M. Franke, F. Marchini, N. Jux, H.-P. Steinrück, O. Lytken, F. J. Williams, *Chem. Eur. J.* **2016**, *22*, 8520-8524.

[238]   C. Schißler, E. K. Schneider, S. Lebedkin, P. Weis, G. Niedner-Schatteburg, M. M. Kappes, S. Bräse, *Chem. Eur. J.* **2021**, *27*, 15202-15208.

[239]   S. A. Vail, D. I. Schuster, D. M. Guldi, M. Isosomppi, N. Tkachenko, H. Lemmetyinen, A. Palkar, L. Echegoyen, X. Chen, J. Z. H. Zhang, *J. Phys. Chem. B* **2006**, *110*, 14155-14166.

[240]   R. Pudi, C. Rodríguez-Seco, A. Vidal-Ferran, P. Ballester, E. Palomares, *Eur. J. Org. Chem.* **2018**, *2018*, 2064-2070.

[241]   M. Meot-Ner, A. D. Adler, *J. Am. Chem. Soc.* **1975**, *97*, 5107-5111.

[242]   P. G. Seybold, M. Gouterman, *J. Mol. Spectrosc.* **1969**, *31*, 1-13.

[243]   M. Gouterman, G. H. Wagnière, L. C. Snyder, *J. Mol. Spectrosc.* **1963**, *11*, 108-127.

[244]   W. Zheng, N. Shan, L. Yu, X. Wang, *Dyes Pigm.* **2008**, *77*, 153-157.

[245] Y. Matt, I. Wessely, L. Gramespacher, M. Tsotsalas, S. Bräse, *Eur. J. Org. Chem.* **2021**, *2021*, 239-245.

[246] H. Muniyasamy, K. Muthusamy, M. Chinnamadhaiyan, M. Sepperumal, S. Ayyanar, M. Selvaraj, *Energy Fuels* **2022**, *36*, 3909-3919.

[247] Z. Zhou, X. Zhang, S. Ma, C. Liu, X. Liu, R. Ghadari, M. Mateen, Y. Yang, Y. Ding, M. Cai, S. Dai, *J. Phys. Chem. C* **2020**, *124*, 2886-2894.

[248] G. Gritzner, J. Kuta, *Pure Appl. Chem.* **1984**, *56*, 461-466.

[249] P. Pracht, F. Bohle, S. Grimme, *Phys. Chem. Chem. Phys.* **2020**, *22*, 7169-7192.

[250] S. Spicher, S. Grimme, *Angew. Chem. Int. Ed.* **2020**, *59*, 15665-15673.

[251] RDKit:Open-source cheminformatics. https://www.rdkit.org

[252] N. M. O'Boyle, M. Banck, C. A. James, C. Morley, T. Vandermeersch, G. R. Hutchison, *J. Cheminform.* **2011**, *3*, 33.

[253] B. Hourahine, B. Aradi, V. Blum, F. Bonafé, A. Buccheri, C. Camacho, C. Cevallos, M. Y. Deshaye, T. Dumitrică, A. Dominguez, S. Ehlert, M. Elstner, T. van der Heide, J. Hermann, S. Irle, J. J. Kranz, C. Köhler, T. Kowalczyk, T. Kubař, I. S. Lee, V. Lutsker, R. J. Maurer, S. K. Min, I. Mitchell, C. Negre, T. A. Niehaus, A. M. N. Niklasson, A. J. Page, A. Pecchia, G. Penazzi, M. P. Persson, J. Řezáč, C. G. Sánchez, M. Sternberg, M. Stöhr, F. Stuckenberg, A. Tkatchenko, V. W. z. Yu, T. Frauenheim, *J. Chem. Phys.* **2020**, *152*, 124101.

[254] S. Grimme, J. Antony, S. Ehrlich, H. Krieg, *J. Chem. Phys.* **2010**, *132*, 154104.

[255] S. Grimme, S. Ehrlich, L. Goerigk, *J. Comput. Chem.* **2011**, *32*, 1456-1465.

[256] TURBOMOLE V7.2 2017, a development of University of Karlsruhe and Forschungszentrum Karlsruhe GmbH, 1989-2007, TURBOMOLE GmbH, since 2007; available from http://www.turbomole.com.

8 APPENDIX

8.1 SUPPLEMENTARY DATA

Thermogravimetric analysis

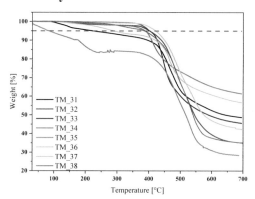

Figure 62: TGA analysis of di-substituted [2.2]paracyclophanes.

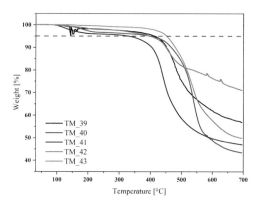

Figure 63: TGA analysis of tetra-substituted [2.2]paracyclophanes.

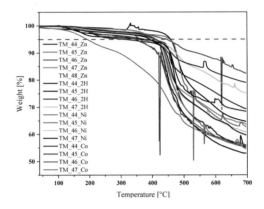

Figure 64: TGA analysis of meso-tetrakis-substituted porphyrins.

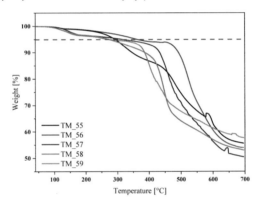

Figure 65: TGA analysis of tetra-substituted TPEs.

8.2 CURRICULUM VITAE

Steffen Otterbach

Born 20.01.1995 in Lauchhammer, Germany

Werderstraße 71A, 76137 Karlsruhe, Germany

Steffen.Otterbach@cdi.eu — +49 1758934241

Education

Collège des Ingénieurs (CDI)	01/2020 – present
MBA Fellow	Munich, Germany
Science and Management Program	
Karlsruhe Institute of Technology (KIT)	01/2020 – present
Ph.D. Candidate in Organic Chemistry (Prof. Dr. Bräse)	Karlsruhe, Germany
Karlsruhe Institute of Technology (KIT)	10/2017 – 09/2019
M.Sc. Chemistry (1.0, with distinction)	Karlsruhe, Germany
Master thesis (Prof. Dr. Bräse)	
"Synthesis of Multifunctional Aromatic Aldehydes and Their Application in Material Science" (1.0)	
Karlsruhe Institute of Technology (KIT)	10.2014 – 09/2017
B.Sc. Chemistry (1.4)	Karlsruhe, Germany
Bachelor Thesis (Prof. Dr. Kleist)	
"Influence of Temperature and Preparation Method on the Synthesis of Bimetallic MCuBTC Materials (M = Fe, Cu)" (1.0)	
Werner-Heisenberg Gymnasium	29.03.2014
Abitur (1.6)	Bad Dürkheim, Germany

Achievements

Ph.D. Scholarship from the German Federal Environmental Foundation (DBU)	01/2021 – present Osnabrück, Germany
Member of the Doctoral College "Energy Transition" of the DBU	01/2021 – present Osnabrück/Berlin, Germany
Jánoz-Rétey-Prize for outstanding accomplishments in Organic Chemistry during the Master Studies	12/2020 Karlsruhe, Germany

8.3 LIST OF PUBLICATIONS

Peer-reviewed Publications:

A. M. Eisterhold, T. Puangsamlee, **S. Otterbach**, S. Bräse, P. Weis, X. Wang, K. V. Kutonova, O. Š. Miljanić, *Org. Lett.* **2021**, 23, 3, 781–785.

J. Bitzer, **S. Otterbach**, K. Thangavel, A. Kultaeva, R. Schmid, A. Pöppl, W. Kleist, *Chem. Eur. J.* **2020**, 26, 5667-5675.

Manuscripts in preparation:

S. Otterbach, D. Elsing, A. Schulz, H. Röhm, M. Kozlowska, W. Wenzel, A. Colsmann, S. Bräse "Novel pseudo-*para* substituted [2.2]Paracyclophanes *via* CH Activation"

S. Otterbach, A. Schulz, D. Elsing, M. Kozlowska, H. Röhm, W. Wenzel, A. Colsmann, S. Bräse "Novel Tetra-substituted [2.2]Paracyclophanes as Hole Transport Materials in Perovskite Solar Cells"

S. Otterbach, M. Krstić, W. Wenzel, C. Rockstuhl, K.V. Kutonova, S. Bräse "Derivatizing of Dimeric BODIPY: Introducing Functional Groups to the Biphenyl Bridge"

Conference Posters:

S. Otterbach, H. Tappert, S. Bräse "Design and Synthesis of Novel Hole Transport Materials for Emerging Active Layers", from May 08th – 13th, 2022, Honolulu, Hawai'i, USA, 2022 MRS Spring Meeting & Exhibit

8.4 ACKNOWLEDGMENTS

Firstly. I would like to thank my doctoral supervisor Prof. Dr. Stefan Bräse for giving me the chance to prepare this work in his group. His support and guidance helped me transition from my masters project to the Ph.D. project without feeling overwhelmed while providing me with the freedom to explore, attempt and learn. His help during my applications for the CDI MBA program, the DBU scholarship and the MRS Spring Meeting proved to be vital as all three applications led to the desired results.

I would like to express my thanks to my collaborators during the last three years. Without the KERASOLAR project funded by CARL ZEISS STIFTUNG I would not hav been able to find my collaborators and obtain the results presented. My thanks goes to the entire team of KERASOLAR for the interesting and interdisciplinary meetings and to my close collaborators especially. These are Alexander Schulz and Dr. Holger Röhm from LTI who helped me during the development of HTMs and have waited patiently until applicable materials were synthesized. Everytime I synthesized new materials I could be sure that they would do everything possible to achieve new results. Next, I would like to thank David Elsing, Dr. Mariana Kozlowska and Prof. Dr. Wolfgang Wenzel for their help regarding theoretical insights into my materials. The meetings have always generated new ideas and visions and I could rely on them to provide me with results. Prof. Dr. Uli Lemmer is kindly acknowledged for the acceptance of the co-reference of this thesis. I would like to thank PD Dr. Patrick Weis for not only helping me with the mass analysis during the cyclotetrabenzoin project which resulted in a publication but also for his fast help and interest in the porphyrin structures when the IOC analytics could not help me any further. I would like to thank Dr. Ksenia Kutonova for her insights and her support at the beginning of my Ph.D. and for everything I could learn from her during my Vertiefer and my Master theses.

I would like to express my gratitude to the German Federal Environmental Foundation (DBU) for offering me a scholarship for my Ph.D. work. I would like to especially thank Dr. Katrin Anneser and Dr. Volker Wachendörfer for inviting me to the Doctoral College "Energy transition" and giving me the chance to meet many like-minded and motivated peers first online and then at the "DBU Salon: Energiewende" in Osnabrück. The yearly seminar weeks have provided me with more insights into the interdisciplinary work regarding nature and energy research, helped me free my head from the lab work, and I am glad to become a member in this alumni network.

Another thanks goes to the CDI and the Science & Mangement program which provided me with great memories of the online- and offline courses. I met many great characters there and I am looking forward to fully embrace the program during my full-time stay.

I would like to thank all the student that I had the chance of supervising in the last years: Tamara Meyer, Eileen List and Barbara Thiele as OCF students; Tamara Meyer, Aulona Sahiti and Pit Turpel as HiWis and Stephan Behling and Oliver Lampert as bachelor students. You have all helped me a lot during my thesis and I wish you all the best for your future.

Of course, this thesis would have probably been very different if I could not have worked on it in lab 306. Thank you for the unforgettable memories and the way things were handled during Covid. I would like to especially thank Dr. Christoph Zippel, Dr. Mareen Stahlberger, Dr. Nicolai Rosenbaum, Jannik Schlindwein and Jens Hohmann for their insights, their practical help, and the good times during the last years. Some puzzles I might have not been able to solve without you.

Fortunately, the unforgettable memories do not end with 306. A big thank you to everyone in the third floor who have made the last 4 years pass so fast and create new friendships. Thank you to Henrik Tappert, Hannes Kühner, Clara Adam, Céline Leonhardt, Dr. Gloria Hong, Jasmin Seibert and Sebastian Bichelmaier for the great time at the MRS Spring Meeting in Hawai'i and making it feel more like vacation than a conference. Thank you to Annalena Leistner, Lukas Langer, Dr. Christoph Grathwol and Lisa-Lou Gracia for embracing bouldering, creating great memories and I am looking forward to solve more projects with you.

A special thank you goes to Jonas Braun, Lara Faden, Simon Oßwald and Jonas Strecker, who have accompanied me on the academic journey from the first semester on and I am glad and thankful to have spent so much time with you. I wish you all the best for your upcoming theses and that we do not loose sight of each other.

This thesis would not have been possible this way without the thorough work of Lukas Langer, Simon Oßwald and Henrik Tappert who took the time to correct and discuss the thesis and I am thankful for their insightful and straightforward comments. The thesis would not look the same without your help!

Thank you to all of AK Bräse for the last years and the friendly work environment. Thank you especially to Janine Bolz and Christiane Lampert for the help with bureaucracy, the scholarship funding and many other small issues. Thank you to FC Bräselona for the good times and thank you to everyone at Campus North for the help during the CV and TGA measurements and for the open communication regarding ELN and Repository.

A big thank you also goes to the IOC team, especially to Angelika Mösle, Lara Hirsch, Despina Savvidou, Lisa Wanner and Jens Hohmann for measuring all my compounds.

Lastly, I would like to thank my family and my friends for their support throughout the entire phase of my Ph.D. and I don't know how it would have gone without you.